International Workshop on Nanotechnology (IWN 2020)

IOP Conference Series: Materials Science and Engineering
Volume 1095

Al-Fallujah, Iraq
24 - 26 December 2020

ISBN: 978-1-7138-3351-2
ISSN: 1757-8981

Printed from e-media with permission by:

Curran Associates, Inc.
57 Morehouse Lane
Red Hook, NY 12571

Some format issues inherent in the e-media version may also appear in this print version.

Institute of Physics
Dirac House, Temple Back
Bristol BS1 6BE UK

Phone: 44 1 17 929 7481
Fax: 44 1 17 920 0979

techtracking@iop.org

Additional copies of this publication are available from:

Curran Associates, Inc.
57 Morehouse Lane
Red Hook, NY 12571 USA
Phone: 845-758-0400
Fax: 845-758-2633
Email: curran@proceedings.com
Web: www.proceedings.com

TABLE OF CONTENTS

International Workshop on Nanotechnology (IWN)

Join with

1st International Conference on Computing and Emerging Sciences

(ICCES 2020)

Erbil, Iraq

24-26 December, 2020

Organized by

Al-Huda University College, Iraq

Conference Chairs

Omer K. Jasim Mohammad, University of Fallujah, **Iraq**

Safia Abbas, Princess Nourah Abdulrahman University (PNU), **KSA**

IOP Conf. Series: Materials Science and Engineering 1095 (2021) 011001 doi:10.1088/1757-899X/1095/1/011001

1st International Conference on Computing and Emerging Sciences

Preface

Al-Huda University College (HUC) is an Iraqi accredited private educational institution. It was founded in 2020.

HUC has launched multiple community reach out initiatives in coincidence with post-war stabilization period. From women empowerment, environment recovery, demining, public health awareness to job creation projects and enhancing local institutional performance, HUC has been a fighter in advancing the local life.

Since its founding, HUC has been offering quality education to multiple rising generations who contribute to an advanced Iraq.

It offers rigorous quality undergraduate program in accounting, Oil engineer Techniques and laboratory Techniques.

Ahead of this international event, HUC administration and board of trustees in addition to International Workshop on Nanotechnology Applications steering committee and chairmanship would like to extend their innermost appreciation for the international researchers, key-note speakers, academicians and experts who are making this event a successful reality. Due to growing concerns about COVID-19, IWN 20 has been provided fully virtual platform, ZOOM, as a another solution . This unfortunate event does give us the opportunity to innovate on how to host an effective remote workshop.

Conference Chair
Omer K, Jasim Mohammad
University of Fallujah

Conference Chair
Safia Abbas
Princess Nourah University

Dec. 24, 2020

IOP Conf. Series: Materials Science and Engineering 1095 (2021) 011001 doi:10.1088/1757-899X/1095/1/011001

1st International Conference on Computing and Emerging Sciences

General Chair Message

With great pride and pleasure, Al-Huda University College announces the First International Workshop on Nanotechnology applications that is joint with 1st International Conference on Computing and Emerging Sciences (ICCES20). The workshop aims to disseminate knowledge and findings among re-searchers worldwide. As new horizons appear, demand for collaboration and discussion increases. The workshop will open novel gateways for all scientists and researchers to share and discuss recent trends in the area.

The goal of IWN '20 is to gather scholars from all over the world to present advances in the relevant fields and to foster an environment conducive to exchanging ideas and information. This WORKSHOP will also provide an ideal environment to develop new collaborations and meet experts on the fundamentals, applications, and products of the mentioned fields.

The Huda University College seizes this opportunity to extend their finest appreciations to IOP and Scopus and all publication sponsors who helped turn the idea of this event into a concrete reality.

Last but most, sincere thanks are due to be extended to Prof.Dr. Badrul Khan, Prof. Rania Hodhod, Professor Maher El-Kady and Dr Amer S. El-ameer for their keen support and seamless cooperation, weathering all challenges to attend and shape this international event.

Conference Chair

Omer K. Jasim Mohammad

IOP Conf. Series: Materials Science and Engineering 1095 (2021) 011001 doi:10.1088/1757-899X/1095/1/011001

1st International Conference on Computing and Emerging Sciences

Organizing Committee Message

Dear Colleagues,

On behalf of the ICCES'20 organizing committee, I am honored and delighted to welcome you to the 1st International Conference on Computing and Emerging Sciences, Erbil, Iraq. I believe we have chosen a venue that guarantees a successful technical conference amid the culture and scenery of Iraq.

Our technical program is rich and varied, it includes 4 keynote speeches and 2 invited talks and around 51 technical papers split between three parallel oral sessions for each day. Besides, there are three co-located workshops and one event on technological entrepreneurship.

In addition to being an excellent forum for information sharing, it will also provide a good opportunity to connect with people working in the field of medical technology in the region. We guarantee that you will find this conference enriched and stimulating.

We would like to encourage you to participate in this most prestigious event and the next version of ICCES. It is only with your contribution that we can ensure the success of our international event.

We look forward to welcoming you in NEXT VERSION at Ramadi, Anbar, Iraq.

Chair of Organizing Committee

Hameed M. Mohammed (*Dean*)

SCITEPRESS DIGITAL LIBRARY

Scopus

IOP Conf. Series: Materials Science and Engineering 1095 (2021) 011001 doi:10.1088/1757-899X/1095/1/011001

1st International Conference on Computing and Emerging Sciences

Program Committee Message

First of all, let me express my profound gratitude for your participation in our international event. ICCES will take virtually place on 24-26 December 2020, organized by the Al-Huda University College.

We have proceeded to hold regular meetings of scientists and researchers involved in information science and related disciplines. Your articles indicate that the continued interest in the detailed procedures has many supporters.

Topics covered in your papers, indicate an unyielding fact that information and technology transcend all sciences,

Also, three special issues contain the selected weighted papers, accepted for publication, which are presented at the ICCES20.

Thank you so much for supporting our event.

Program Chair

Prof. Rania Hodhod

Columbus State University

1st International Conference on Computing and Emerging Sciences

SCITEPRESS DIGITAL LIBRARY

Scopus

SCITEPRESS DIGITAL LIBRARY

iCCS

Alhuda
AL HUDA UNIVERSITY COLLEGE

International Workshop on Nanotechnology (IWN)

join with 1st International Conference on Computing and Emerging Sciences

24-26 December، 2020

IOP Conf. Series: Materials Science and Engineering 1095 (2021) 011003 doi:10.1088/1757-899X/1095/1/011003

Peer review declaration

All papers published in this volume of *IOP Conference Series: Materials Science and Engineering* have been peer reviewed through processes administered by the Editors. Reviews were conducted by expert referees to the professional and scientific standards expected of a proceedings journal published by IOP Publishing.

- **Type of peer review: Triple-blind**
- **Describe criteria used by Reviewers when accepting/declining papers. Was there the opportunity to resubmit articles after revisions?**
- **The reviewer criteria listed briefly as below:**
  - *Author submit paper to the conference.*
  - *The workshop secretary received the paper and send it to the member of Scientific Committee (SC).*
  - *SC member check the topic and test the plagiarism.*
  - *Eliminate the name of authors from the title page and forward to three reviewers.*
  - *Given 30 days for each one, in order to complete the reviewer form.*
  - *I f two reviewers decline the paper, SC member directly inform the workshop secretary to send notification to the author, else, paper should be revised as reviewers comments if two of them accepted.*

- **Conference submission management system:**
  - *Easy-chair*

- **Number of submissions received:**
  - *44 papers*

- **Number of submissions sent for review:**
  - *44 papers*

- **Number of submissions accepted:**
  - *15 papers*

- **Acceptance Rate (Number of Submissions Accepted / Number of Submissions Received X 100):**
  - *- 29.9%*

- **Average number of reviews per paper:**
  - **Three reviewers**

- **Total number of reviewers involved:**
  - *134 from 23 countries*

- **Any additional info on review process (ie plagiarism check system):**
  - *Turnitin solutions*

IOP Conf. Series: Materials Science and Engineering 1095 (2021) 011003 doi:10.1088/1757-899X/1095/1/011003

- **Contact person for queries:**
  - Prof. Omer K. Jasim, Head of QA at University of Fallujah.
  - Omerk.jasim@uofallujah.edu.iq
  - 00964-7812182165

A Comparative Study of Performance Between Two Combined Ventilation Systems and Their Effect On Indoor Air Quality and Thermal Comfort Inside Office Rooms

Hussien Aziz Saheb[1], Ala'a Abbas Mahdi [2] and Qusay Rasheed Al-amir [3]

[1] College of Engineering, Babylon University, Iraq - E-mail: hassan1245konh@gmail.com
[2] College of Engineering, Babylon University, Iraq - E-mail: alaa.mahdi1959@gmail.com
[3] College of Engineering, Babylon University, Iraq - E-mail: qusay1972@gmail.com

[1]**Corresponding author:** hassan1245konh@gmail.com

Abstract. The most important problems that must be solved and improved in ventilation research are indoor air quality and thermal comfort. The Iraqi climate is characterized by being very hot and dry, so fresh cold air must be provided to the occupants of offices, factories, workshops and others to improve thermal comfort and increase their efficiency. This study focused on improving the quality of inhaled air by using two types of ventilation systems (mixing and displacement ventilation) in combined with a personal ventilation system. This study was a set of numerical tests that were conducted to predict the temperature distribution, air movement and speed inside a closed office room subject to the Iraqi climate in terms of weather conditions. This numerical study was carried out using (AIRPAK3.0.16) which is used to solve turbulence equations, Naiver stock, energy equations, and use of (FVM). A thermally isolated room was simulated with a personal ventilation system that supplies air at temperatures from 21 to 23 °C for case I (personal and mixing) and supplies air at a temperature from 18 to 20 °C for case II (personal and displacement) the air velocity for both cases was 0.6 m/s. It was found that there was a positive effect of the personal ventilation system in both cases, but its effect was more pronounced in the first case, through numerical calculations for each of the effectiveness of heat removal (Ɛt) and air distribution performance index (ADPI), where (74.361) and (1.549) were found for the first case, and (68.321) and (1.71) for the second case.

Keywords: Thermal comfort, AIRPAK, Computational fluid- dynamics (CFD), Mixing ventilation (MV), Displacement ventilation(DV), Indoor air quality (IAQ). , Personalized ventilation (PV)

1. Introduction

Nowadays, with industrial progress and technological development, most people spend 80% of their time in closed buildings and environments full of pollutants and emissions, so the need for innovations for ventilation systems has emerged. It is very necessary to have indoor environments that are distinguished by providing thermal comfort to their occupants and are suitable in terms of recreation and health for them. Recent research on building ventilation issues has focused on both thermal comfort and indoor air quality, which are considered among the most important factors affecting the quality of the indoor environment and human comfort.

There are many types of mechanical ventilation systems, but there are two types, the most common, which are mixing ventilation and displacement ventilation [1]. Mixing ventilation, It is a ventilation system that mixes the supply air with the old air of the room so that the air inside the room becomes at the same design temperature. In the cooling mode, the supply air is provided at a temperature of 13 °C. Displacement ventilation system (DV)is defined as the displacement of the room's old air and its replacement with new air coming from the outside of the room [2]. Many problems arise when using ventilation systems, so solutions must be found to them as one of the most common of these problems is the mixing of the cold and fresh air supplies coming from the diffusers with the hot room air full of pollutants before it reaches the breathing zone[3]. Hence the need to use personal ventilation systems, which supply fresh cold air directly to the breathing zone. Personal ventilation systems increase the quality of inhaled air by continually preparing and supplying the breathing zone with fresh, cool air[4]. Personal ventilation systems (PV) are preventive systems as they work to protect people from airborne diseases, and cold air supplies from personal ventilation systems reduce the percentage of pollutants in the breathing area, and this feature is one of the most important advantages of personal ventilation systems. [5]. Personal ventilation systems (PV) were used in addition to mixing ventilation systems, and displacement ventilation systems because personal ventilation systems are distinguished by their ability to give full freedom to occupants of the space by controlling the surrounding climate. [6]. The addition of personal ventilation systems inside buildings improves thermal comfort, increases and improves indoor air quality, as well as reduces pollutants and eliminates symptoms of SBS [7, 8, 9].

2. Related Work

Al-ssaad et al., 2017 [10] investigated how to improve mechanical ventilation systems by adding and combining a personal ventilation system and a mixed ventilation system. This study was conducted numerically and experimentally in laboratories under standard conditions. And it was found that when using the personal ventilation system, the thermal comfort was increased and it was about 0.95, as for the efficiency of ventilation it was 77%, and the use of a personal ventilation system led to a decrease in energy costs by about 21.34%.

Al-ssaad et al., 2018 [11] studied the effectiveness of adding a personal ventilation system to a mixing ventilation system and achieving the best performance. This study was simulated with available numerical methods to predict air flows at a suitable supply frequency for space occupants. This study was conducted under certain conditions where the room air temperature was considered to be 25 °C, while the air temperature for supplying personal ventilation systems was 22 °C. This study found that by increasing the frequency at a constant flow rate, the thermal comfort will increase by about 15.2%, and when the average flow rate increases at a constant frequency, the thermal comfort will decrease at a low frequency of 0.3 Hz, but it remains acceptable at higher. 0.5 Hz frequency.

Yakoob et al., 2019 [12] investigated increasing and improving the efficiency of air exchange inside office rooms by using the displacement ventilation system with the mixing ventilation system. This study was conducted by experimental methods under a laboratory, as well as the distribution of air temperatures and air flows by numerical methods. The RNG disturbance model was adopted. This study has demonstrated, that the presence of a combined system of personal ventilation systems with displacement ventilation will lead to an increase in the thermal comfort of a person, depending on the values of (Ɛt) and (ADPI) which are (71%) and (1.8), respectively. Using a personal ventilation system with a displacement ventilation system will increase the air quality and thermal comfort as well as increase the quality of the air inhaled in the breathing zone.

Most of the studies in the past and present time focused on studying air movement and temperature distribution inside the room, as well as the locations of diffusers and their geometric shapes.

This study took another side as it focused on the following

- This study aims to find out which ventilation system is suitable for use with a personal ventilation system inside office rooms.
- Study how to increase the quality of inhaled air and thermal comfort by choosing the most appropriate system inside the office rooms.
- Investigating both indoor air quality and thermal comfort by studying and analyzing the values of two important indicators, namely air diffusion performance index(ADPI) and heat removal effectiveness (Ɛt).

In this numerical work, it was used the ANSYS Fluent AIRPAK 3.0.16 for the purpose of simulating a thermally isolated room and predicting the air flows inside it, as well as simulating the temperature distribution inside the room and around its occupants.

3. Description Room Configuration and Assumptions

An office room was simulated by numerical methods and its dimensions were (3×2.5×2.5) m as shown in Fig (1), and the conditions under which the study was conducted were where the 3D computational model of the steady-state. The room was numerically simulated under the Iraqi climate and the room was thermally insulated. Two types of ventilation systems were simulated, which are the mixing ventilation system and the displacement ventilation system, as shown in Fig (1). The personal ventilation system supplies air at a temperature of 22 °C at a speed of 0.6 m / s at a distance of 0.5 m from the face of the simulated seated person. A person sitting in front of a computer is simulated and there is a personal ventilation system in front of the sitting person that draws air from a height of 0.4 m from the floor of the room and supplies it at a height of 1.1 m. The mixing ventilation system supplies the air at a speed of 2.5 m/s from a diffuser just below the ceiling onto the north wall of the room. As for the displacement ventilation system, the air diffuser installed near the floor of the room supplies the air at a speed of 0.25 m/s. The heat gain (only sensible heat) for a person is 75 W, computer 60 W, and lump 100 W. It was used (Cooling Load Temperature Difference Method (CLTD)) [13].

4. Computational Fluid Modeling

The governing equations in this numerical study are [14].

4.1 Governing Equations:

$$\frac{\partial}{\partial x}(\rho u)+\frac{\partial}{\partial y}(\rho v)+\frac{\partial}{\partial z}(\rho w)=0 \quad (1)$$

$$\frac{\partial}{\partial x}(\rho uu)+\frac{\partial}{\partial y}(\rho uv)+\frac{\partial}{\partial z}(\rho uw)=-\frac{\partial p}{\partial x}+\frac{\partial}{\partial x}\left(\mu\frac{\partial u}{\partial x}\right)+\frac{\partial}{\partial y}\left(\mu\frac{\partial u}{\partial y}\right)+\frac{\partial}{\partial z}$$

$$\left(\mu\frac{\partial u}{\partial z}\right)+\frac{1}{3}\frac{\partial}{\partial x}\left[\mu\left(\frac{\partial u}{\partial x}+\frac{\partial v}{\partial y}+\frac{\partial w}{\partial z}\right)\right]+\frac{\partial}{\partial x}\left(-\rho\overline{u'u'}\right)+\frac{\partial}{\partial y}\left(-\rho\overline{u'v'}\right)+\frac{\partial}{\partial z}\left(-\rho\overline{u'w'}\right)+\rho g_x \quad (2)$$

$$\frac{\partial}{\partial x}(\rho uv)+\frac{\partial}{\partial y}(\rho vv)+\frac{\partial}{\partial z}(\rho vw)=-\frac{\partial p}{\partial y}+\frac{\partial}{\partial x}\left(\mu\frac{\partial v}{\partial x}\right)+\frac{\partial}{\partial y}\left(\mu\frac{\partial v}{\partial y}\right)+\frac{\partial}{\partial z}$$

$$\left(\mu\frac{\partial v}{\partial z}\right)+\frac{1}{3}\frac{\partial}{\partial y}\left[\mu\left(\frac{\partial u}{\partial x}+\frac{\partial v}{\partial y}+\frac{\partial w}{\partial z}\right)\right]+\frac{\partial}{\partial x}\left(-\rho\overline{u'v'}\right)+\frac{\partial}{\partial y}\left(-\rho\overline{v'v'}\right)+\frac{\partial}{\partial z}\left(-\rho\overline{v'w'}\right)+\rho g_y \quad (3)$$

$$\frac{\partial}{\partial x}(\rho uw)+\frac{\partial}{\partial y}(\rho vw)+\frac{\partial}{\partial z}(\rho ww)=-\frac{\partial p}{\partial z}+\frac{\partial}{\partial x}\left(\mu\frac{\partial w}{\partial x}\right)+\frac{\partial}{\partial y}\left(\mu\frac{\partial w}{\partial y}\right)+\frac{\partial}{\partial z}\left(\mu\frac{\partial w}{\partial z}\right)+\frac{1}{3}\frac{\partial}{\partial z}\left[\mu\left(\frac{\partial u}{\partial x}+\frac{\partial v}{\partial y}+\frac{\partial w}{\partial z}\right)\right]+$$

$$\frac{\partial}{\partial x}\left(-\rho\overline{u'w'}\right)+\frac{\partial}{\partial y}\left(-\rho\overline{v'w'}\right)+\frac{\partial}{\partial z}\left(-\rho\overline{w'w'}\right)+\rho g_z \quad (4)$$

$$\frac{\partial}{\partial x}(\rho ut)+\frac{\partial}{\partial y}(\rho vt)+\frac{\partial}{\partial z}(\rho wt)=\frac{\partial}{\partial x}\left(\Gamma\frac{\partial t}{\partial x}\right)+\frac{\partial}{\partial y}\left(\Gamma\frac{\partial t}{\partial y}\right)+\frac{\partial}{\partial z}\left(\Gamma\frac{\partial t}{\partial z}\right)+\frac{\partial}{\partial x}\left(-\rho\overline{u't'}\right)+\frac{\partial}{\partial y}\left(-\rho\overline{v't'}\right)+\frac{\partial}{\partial z}\left(-\rho\overline{w't'}\right)+$$

$$S_t \quad (5)$$

4.2 Models of Turbulence:

The simulation was done with the use of 4 types of disturbance models, and the disturbance model was the best in terms of speed of performance, the accuracy of results, and its proximity to realism was the type disturbance model the turbulent model RNG K-Ɛ. This was fully compatible with Chen, [15] investigated eight K-Ɛ models for mixed convection flow and found that the Renormalization Group (RNG K-Ɛ) model was the best, closest to reality and acceptable results. Yakhot [16]. the Renormalization Group (RNG K-Ɛ) better kind of turbulence models have tested. Yuan, [17] studied predicting the distribution of indoor pollutants in the displacement ventilation chamber using the model (RNG K-Ɛ), the equation for the model of turbulent used are [18]:

$$(\rho U_i)\frac{\partial k}{\partial x_i} = \mu_t S^2 + \frac{\partial}{\partial x_i}\left[\alpha_i \mu_{eff}\frac{\partial k}{\partial x_i}\right] - \rho\varepsilon \tag{6}$$

Where $S \equiv \sqrt{2S_{ij}S_{ij}}$ $S_{ij} = \frac{1}{2}\left(\frac{\partial U_j}{\partial x_i} + \frac{\partial U_i}{\partial x_j}\right)$ (7)

$$\rho u_i \frac{\partial \varepsilon}{\partial x_i} = C_{1\varepsilon}\frac{\varepsilon}{k}\mu_t S^2 + \frac{\partial}{\partial x_j}\left[\alpha_\varepsilon \mu_{eff}\frac{\partial \varepsilon}{\partial x_i}\right] - C_{2\varepsilon}\rho\frac{\varepsilon^2}{k} - R \tag{8}$$

4.3 Numerical Solution and Mesh Generation:

The values of (ε_t) and (ADPI) were calculated through numerical simulation by AIRPAK3.0.16. In this work, a method was used the finite volume method. The turbulence model (RNG, K-ε) used was, after several attempts, to reach the most acceptable results and the lowest error rates. Hexa unstructured geometry to discretize. For this function, all types of the element were used to fit the grid to the geometry. After reaching the error rate 10^{-3}, the results became more acceptable and the solutions were completely stable, as shown in Fig (3). In this numerical simulation (finer mesh) is used to reduce the error rate. After testing several types of mesh, coarser, coarse, medium, and fine type were generated, to perform the test and the most appropriate for this study was the type (coarser). The AIRPAK3.0.16 software used to create the model and meshed study cases based on many testing meshes as shown in Figs (4 and 5), the dimensions of edges of the room in (x, y , z) meshed as interval size of (0.04), mesh parameter normal, with Max side ratio (2).

4.4. Boundary Conditions

Boundary conditions are one of the most important factors that greatly and directly affect the success and accuracy of CFD simulation, and they affect the reliability of results for problems that are solved numerically. The boundary conditions were assumed as consisting of one type of inlet and flow velocity as shown in Table (3). Also, the walls were considered fixed, non-slip, insulating, not allowing heat to transfer through them (thermally insulated. Reaching accurate solutions to the end is impossible, so it is necessary to calculate and determine the residual error acceptable for different conditions such as continuity, speed components, and energy. To ensure the validity of the program and the results in this study, this was verified after comparing the experimental results of another study with the numerical results obtained with this numerical simulation by (AIRPAK3.0.16) where the validity of the RNG K-model was verified as shown in Fig (4) . This comparison gave a convergence between the numerical data of the study and the experimental data [19], where the perturbed RNG K-epsilon model was used, and it was found that the percentage of deviation is about 5.37%.

5. Results and discussion

The results will be discussed numerically in this part of this work, for the two cases in light of the Iraqi climate. The air distribution inside the office room and the temperature distribution inside the room was studied.

- **Case I (Personal and Mixing Ventilation System)**

The contour of the temperature of air (z-plane at z=1.25m) is shown in Fig. (4). This figure shows the entry of air from the diffuser installed in the upper part of the walls, where the air enters at a temperature of 17 °C and then begins to rise due to the thermal loads and heat sources inside the room. Fig. (5) indicates the contour of the speed of air. The air enters at a high velocity of 2.5 m/s mixed with the air in the room and there is a uniformity in the distribution of temperature.

Personal ventilation devices (PV) draw air from a height of 0.4 m. At a height of 0.4 m from the floor of the room, the temperature for case I is within (21-23) °C, which provides suitable cool air to the breathing zone for the occupant of the space as shown in Fig. (6). Therefore, the air is prepared to the breathing zone for the occupant of the space at an appropriate and acceptable temperature as shown in Fig. (7).

- **Case II (Personal and Displacement Ventilation System)**

It is noticed that the temperature distribution in case I is more uniform than the temperature distribution in case II, and as shown in Figs. (8 and 9).

Also, the use of mixed ventilation systems leads to a high velocity of air supply, which allows for uniformity in the velocity of air inside the room. The use of personal ventilation systems does not affect the air currents and its movement within the room, as shown in Fig (9). As for the use of personal ventilation systems with displacement ventilation systems, it was found that there was a non-uniform in the distribution of the velocity of air inside the room, and the temperature was non uniform, as it was the lowest possible on the floor and the highest possible near the ceiling of the room and this affected the thermal comfort and the air quality inside the room.

The contour of the mean age of air is shown in Fig. (10). By calculating the mean age of the air in the breathing zone, it was found that the age of the air in the breathing zone for case I which is shorter than the age of the air in the breathing zone for the case II, as shown in Fig. (11). This means the availability of fresh, cold air as soon as possible, which leads to a decrease in the concentration of pollutants in the breathing area, thus increasing thermal comfort for humans and an increase in the quality of their inhaled air.

The distribution of temperatures inside the laboratory room at six different altitudes, where it is noticed that the temperatures in the case I rise as the height from the floor of the room decreases. As for-case II, the air temperatures rise with the increase in the height from the floor of the room, as it is noticed that the difference in temperatures is for the first case. It is very little, which means a uniform and homogeneous distribution of air. As for the temperature distribution of case II, there will be a clear difference in the temperature difference at different heights from the room floor, and this leads to the mixing ventilation system being considered better in terms of temperature distribution and air velocity inside the room as well as the concentration Pollutants.

The temperatures were calculated in three poles (x,y,z), pole 1(2,0,1.25), pole2(1,0,0.5) and pole 3 (1,0,2)as shown in the Figs. (12). It is noticed that the temperatures are uniform and there is very little difference in temperature at different altitudes from the floor of the room, and this means a uniform distribution of temperatures inside the whole room.

In the case II, as shown in Fig (13) when measuring the temperatures in the three poles, (x,y,z), pole 1(2,0,1.25), pole2(1,0,0.5) and pole 3 (1,0,2) it is noticed that the temperature distribution is less than possible near the floor of the room and increases as the height from the floor increases until it becomes more as possible near the ceiling.

When making a comparison between I and II case, the following was found:

- The mean age of the air inside the room for case II is less than the average age of the air for case II, which means an increase in the supply of cold fresh air continuously for case I.
- In case I, the cold air coming from the diffuser mixes with the room air before it is withdrawn by the personal ventilation systems, which means that the air is supplied from (21-23) °C at a speed suitable for breathing.
- In-case II, the air is supplied from a diffuser near the floor of the room and at a low speed, so mixing of the air does not occur, and the air is drawn in by personal ventilation systems at low temperatures that may be unsuitable for human breathing.
- The effect of using personal ventilation systems is widely seen with the displacement ventilation system because the air velocity of the air diffuser is lower for the case II compared to the air velocity for case I.

6. Conclusions

This numerical work dealt with indoor air quality and how to improve it, using personal ventilation systems with mixing and displacement ventilation systems as two different cases. The most important conclusions of this study are:

1- The use of a personal ventilation system will improve and increase the indoor air quality and thermal comfort of the occupants.

2- The best heat removal effectiveness obtained in case II, while the best Air Diffusion Performance Index (ADPI) obtained in case I.
3- The personal ventilation system is effective for both cases, but the effect is more positive in case I, depending on the temperatures of the air being supplied.
4- The study showed that the use of personal ventilation systems increases the quality of inhaled air in the breathing zone and case I is the best under the same conditions.
5- The limits of thermal comfort found in case I (personal and mixing ventilation), due to depending on the numerical results obtained where the values of (ADPI) and (ε_t) are (74.361) and (1.549) respectively for-case I.

The primary suggestions for future works are as follows:

1- Studying different models and calculating the numerical results of the unsteady state turbulent flow.
2. Studying the conditions of thermal comfort, and the quality of the indoor air for one or more residents in parking and other activities inside the office room (walking, standing, and other activities) with different metabolic rates, as well as thermal insulation of clothing and conditions of relative humidity and others.
3. Studying the different shapes and types of air supply diffusers, such as the circular diffuser, the swirl inlet, and the four-way diffuser in addition to the ceiling diffuser.
4. Studying the effectiveness and capacity of different ATD PV devices to reduce and mitigate the transmission of aerosol disease to healthy occupants in the event of an infected agent in an adjacent office or room.

Nomenclature	***Description***	***Unit***
A	Surface Area for Wall	(m2)
C_p	Specific heat of the air at pressure constant	(kJ/kg.K)
C1ε, C2ε	Coefficient in the specific dissipation rate	
D	Diameter	(m)
dx dy dz	Control volume	(m)
G	Gravitational acceleration	(m/s^2)
N	Total number of draft temperature points measured in occupied zone	
N_θ	Number of points of draft temperature measured in occupied zone	
P	Pressure	(N/ m^2)
Q	Heat transfer through the wall.	(W)
St	Source term for the rate of thermal energy Production	(J/kg)
T	Temperature	(°C)
u,v,w	component of speed in x,y and z-directions	(m/s)
ui,j,k	Speed at cell (i,j,k)	(m/s)
x,y,z	Cartesian coordinates	(m)
Vx	Local air speed	(m/ s)
Vroom	Volume of room (m3)	
Q_s	design flow rate of the supply air	(l/s)
q_{ex}	Cooling load for the heat conduction through the walls and transmitted solar radiation	(W)
q_l	Cooling load for the overhead lighting	(W)
q_{oe}	Cooling load for occupant, desk lamp and equipments	(W)
Q_t	Total heat transfer	(W)
T_{av}	average room temperature	(°C)
Te	Exhaust Air Temperature	(°C)
T_m	outlet design temperature	(°C)
T_o	outlet temperature	(°C)
T_x	local temperature	(°C)
T_{sp}	Setup (design) temperature.	(°C)

Greek Symbols		
ρ	Air density	(kg/m3)
ε	Turbulent energy dissipation rate.	(J/kg .s)
σ	Prandtl or Schmidt number	-
Σk	Model constant	-
Σε	Model constant	-
Ω	Specific dissipation rate	(1/s)
Γ	Diffusion coefficient	(m2/s)
μ_t	Turbulent viscosity	(N.s/m2)
$\dot{\Omega}$	Rotation speed	(rad/s)
ΔThf	The difference in temperature from head to foot level.	(°C)
αk, αε	Coefficient in the specific dissipation rate	-
ε_t	Effectiveness temperature	-
Tt	Turbulent Reynolds stress	-
Sub-Scripts		
av.	Average	
c	Correct	
e	Exhaust Air	
ex	External	
f	Floor	
hf	Head to foot level.	
H	Hydraulic	
i	Inside	
i,j,k	Location of point in a Cartesian grid	
l	Overhead light	
Abbreviations		
ACH	Change of air per hour	
ADPI	Air Distribution Performance Index	
ASHRAE	American Society for Heating, Refrigeration, and Air Conditioning Engineers	
CFD	Computational Fluid Dynamics	
PV	Personalized ventilation	
PMV	Personal-mixing Ventilation	
FVM	Finite Volume Method	
EDT	Effective Draft Temperature	
IAQ	Indoor air quality	
MV	Mixing Ventilation	
RNG	Re-Normalization Group	
SBS	Sick Building Syndrome	
ATD	Air terminal device	
PAQ	Perceived air quality	
DV	Displacement Ventilation	

References

[1] H. B. Awbi, "Ventilation of Buildings", Taylor and Francis group, second edition, 2003.

[2] Garcia, Pablo. "Expanded Impinging Jet Air Supply Method: Air Terminal Device Analysis and Optimization.", 2012.

[3] Melikov, Arsen Krikor. "Personalized ventilation." Indoor air 14, 157-167, 2004..

[4] Kaczmarczyk, Jan, A. Melikov, and Povl Ole Fanger. "Abstract." Indoor Air 14, 17-29, 2004.

[5] Sandberg, Mats, Alan Kabanshi, and Hans Wigö. "Is building ventilation a process of diluting contaminants or delivering clean air?." Indoor and Built Environment 29.6 , 768-774,2020.

[6] Nielsen, P. V., Bartholomaeussen, N. M., Jakubowska, E., Jiang, H., Jonsson, O. T., Krawiecka, K., ... & Soennichsen, M. Chair with integrated personalized ventilation for minimizing cross infection. In The International Conference on Air Distribution in Rooms, Roomvent. FINVAC ry,2007.

[7] Kroner, Walter M. "Environmentally responsive workstations and office-worker productivity." ASHRAE Transactions 100.2, 750-755, 1994.

[8] Chen, Huijuan. Experimental and numerical investigations of a ventilation strategy–impinging jet ventilation for an office environment. Diss. Linköping University Electronic Press, 2014.

[9] Olesen, Bjarne W. "The philosophy behind EN15251: Indoor environmental criteria for design and calculation of energy performance of buildings." Energy and Buildings 39.7, 740-749, 2007.

[10] Al Assaad, D., Ghali, K., Ghaddar, N., & Habchi, C. Mixing ventilation coupled with personalized sinusoidal ventilation: Optimal frequency and flow rate for acceptable air quality. Energy and Buildings, 154, 569-580.2017

[11] Al-Assaad, Douaa, Nesreen Ghaddar, and Kamel Ghali. "Performance of Mixing Ventilation System Coupled With Dynamic Personalized Ventilator for Thermal Comfort." ASME 2017 Heat Transfer Summer Conference. American Society of Mechanical Engineers Digital Collection, 2017.

[12] Asmaa K.Yakoob, Alaa A. Mahdi and Qusay R. Alamir,"Evaluation of Air Exchange Efficiency in Rooms with Personal Ventilation in Conjunction with Displacement Ventilation Systems", International Journal of Mechanical and Mechatronics Engineering IJMME-IJENS, 2019.

[13] Iraqi cooling code, 2012.

[14] H. B. Awbi,. Ventilation systems: design and performance. Routledge, 2007.

[15] Melikov, A. K., Cermak, R., Kovar, O., Forejt, L., Wai, T. K., Sakhar, S. C., & Cheong, D. Impact of airflow interaction on inhaled air quality and transport of contaminants in rooms with personalized and total volume ventilation. In 7th International Conference on Healthy Buildings 2003 (pp. 592-597). Natiol University of Singapore, Department of Buildings.2003

[16] Versteeg, H. K., and W. Malalasekera. "An introduction to computational fluid dynamics." Finite Volume Method, Essex, Longman Scientific & Technical, 1995.

[17] Yuan, X., Chen, Q., Glicksman, L. R., Hu, Y., & Yang, X. Measurements and computations of room airflow with displacement ventilation. Ashrae Transactions, 105, 340.1999.

[18] Yakhot, V. S. A. S. T. B. C. G., et al. "Development of turbulence models for shear flows by a double expansion technique." Physics of Fluids A: Fluid Dynamics 4.7, 1510-1520, 1992.

[19] YAN, Shuai; LI, Xianting. Analytical expression of indoor temperature distribution in generally ventilated room with arbitrary boundary conditions. Energy and Buildings, 208: 109640, 2020.

[20] Ali Aedan, "experimental and numerical investigation of displacement ventilation system with assessment of four turbulence models" M.Sc. Thesis, University of Babylon, Iraq, 2016.

Table 1: Room configuration for (Case-I)

Item	Dimensions (m)			Heat(W)
	X-start	Y-start	Z-start	
Office Room	3	2.5	2.5	
Mixing device	0.05	0.1	0.16	
Exhaust grille	0	0.1	0.12	
Person no.1	0.2	1.1	0.2	75
Computer_no.1	0.3	0.3	0.3	60
Light	0.1	0.05	0.2	100
Computer table1	0.4	0.76	1	

Table 1: Room configuration for (Case-II)

Item	Dimensions (m)			Heat(W)
	X-start	Y-start	Z-start	
Office Room	3	2.5	2.5	
Displacement device	0.05	0.2	0.75	
Exhaust grille	0	0.15	0.56	
Person no.1	0.2	1.1	0.2	75
Computer_no.1	0.3	0.3	0.3	60
Light	0.1	0.05	0.2	100
Computer table1	0.4	0.76	1	

Table 2: The required conditions inside the office buildings in the Iraqi climate during the summer season, [11].

Property / Item	DBT	RH	Recommended air velocity inside the office room
The conditions required within the office buildings	23 °C – 26 °C	40% – 50%	0.25m/s_0.13m/s

Table 3: Boundary Conditions

part	types	Condition of momentum	
		Motion of wall	Shear condition
Person	wall	Stationary	Not Slipping
computer	wall	Stationary	Not Slipping
side walls, floor ceil,	wall	Stationary	Not Slipping
tables and lights	wall	Stationary	Not Slipping
Supply air diffuser	Speed inlet	magnitude, normal to boundary	
Extract Grill	Pressure outlet	-Gauge Pressure = (0 Pa), [constant]. -Backflow direction specification Method: (Normal to Boundary).	

Table 4: Numerical values of ADPI and effectiveness (Ɛt) temp. for each ventilation type

Ventilation devices used	Cases	
CASE	Case I	Case II
ADPI%	74.361	68.321
Effectiveness	1.549	1.71

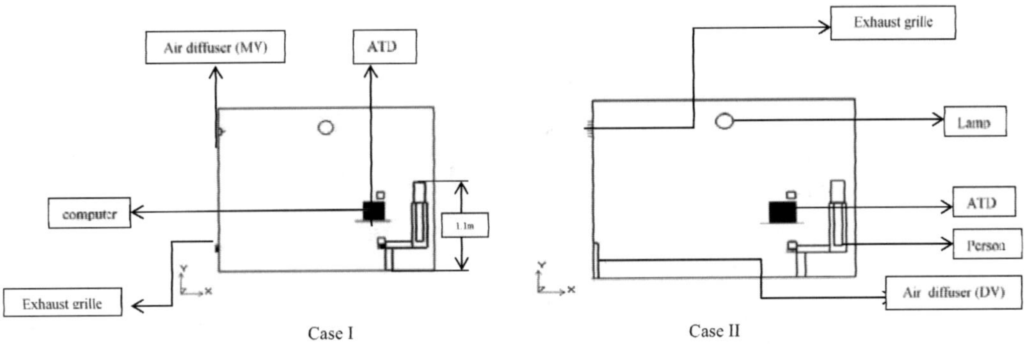

Fig.1. Model of the room

IOP Conf. Series: Materials Science and Engineering 1095 (2021) 012001 doi:10.1088/1757-899X/1095/1/012001

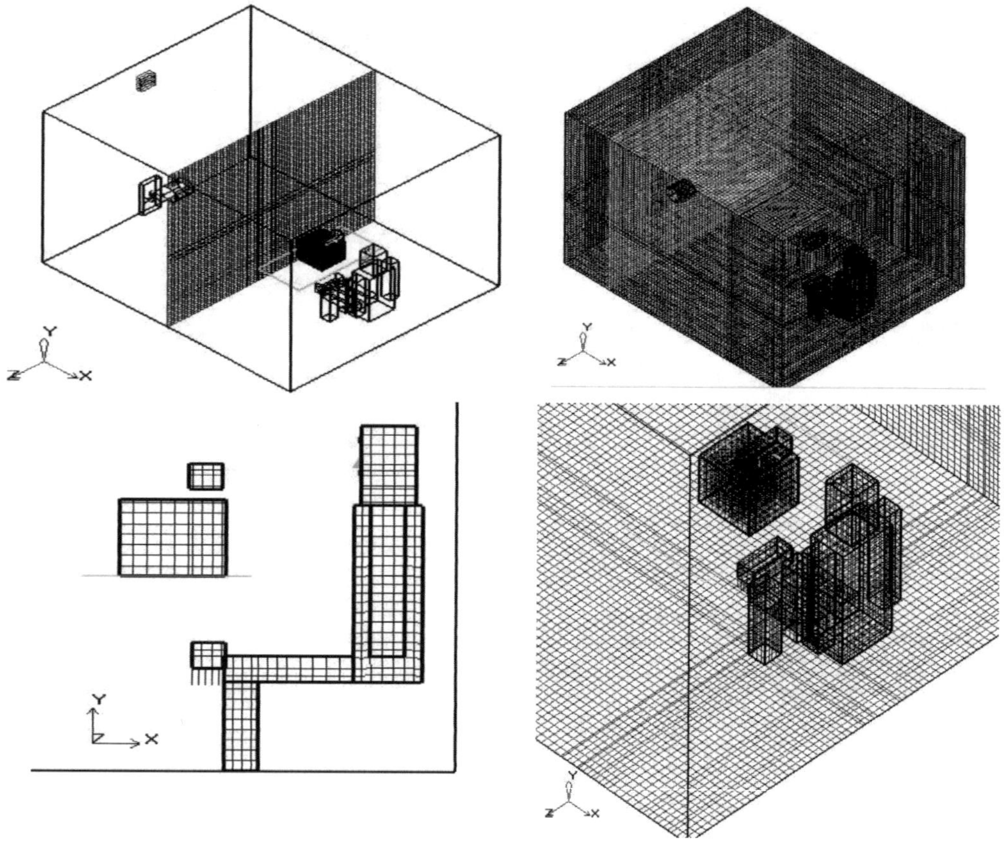

Fig. 2. A part from meshed model

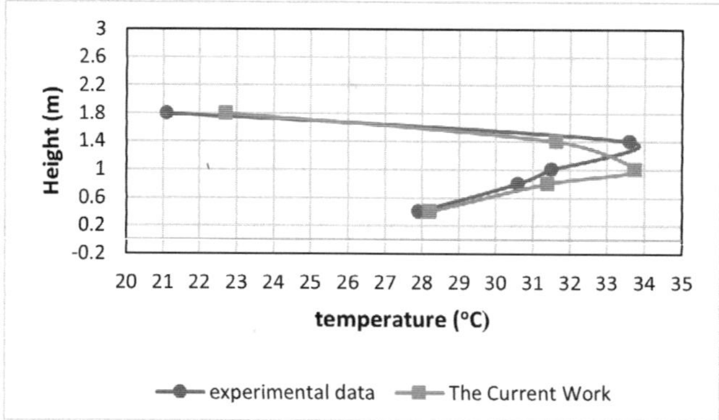

Fig. 3. Validation between the numerical data of this study and an experimental data.[20].

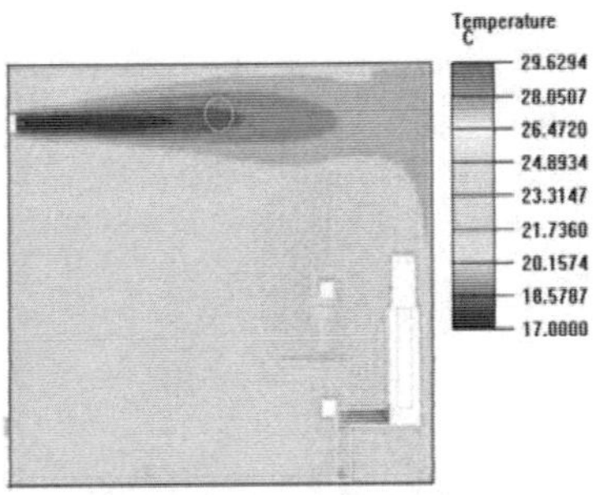

Fig. 4 .Contour of air temperature (z-plane)

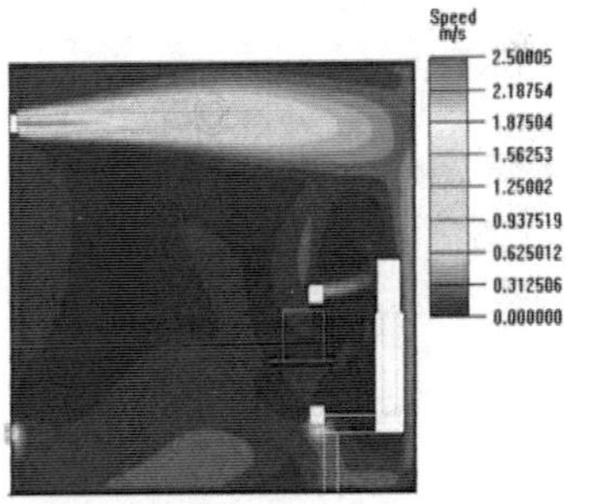

Fig. 5. Contour of air speed (z-plane)

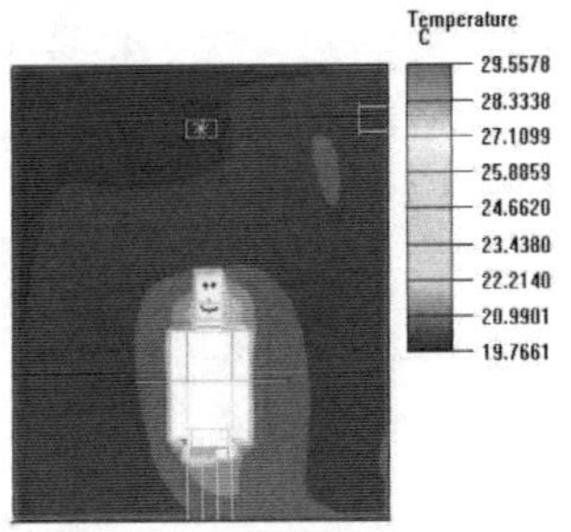

Fig.6. Contour of air temperature (x-plane)

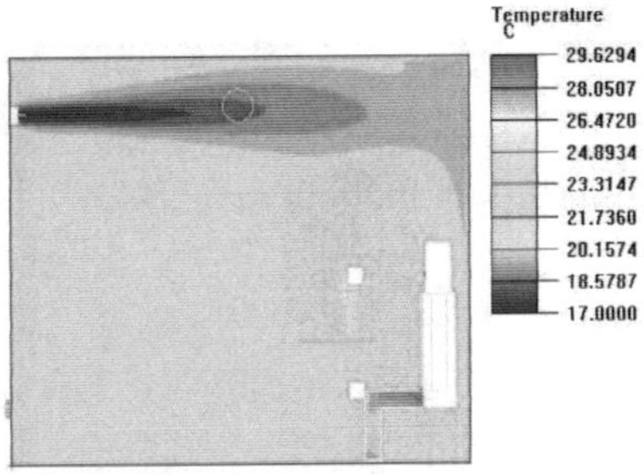

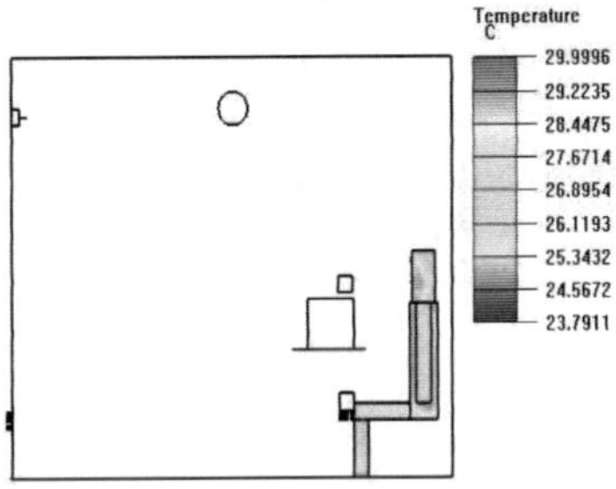

Fig. 7. Contour of air temperature case I (z-plane)

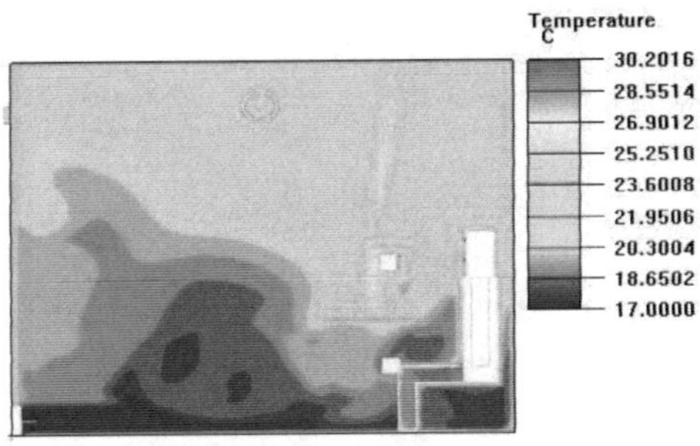

Fig. 8. Contour of air temperature case II (Z-plane)

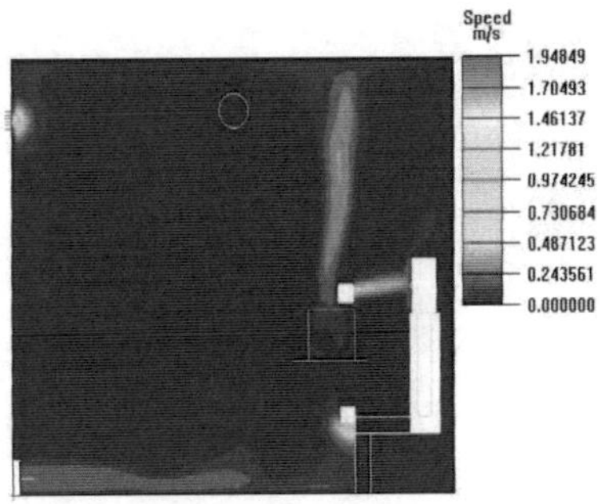

Fig. 9. Contour of air speed case II (z-plane)

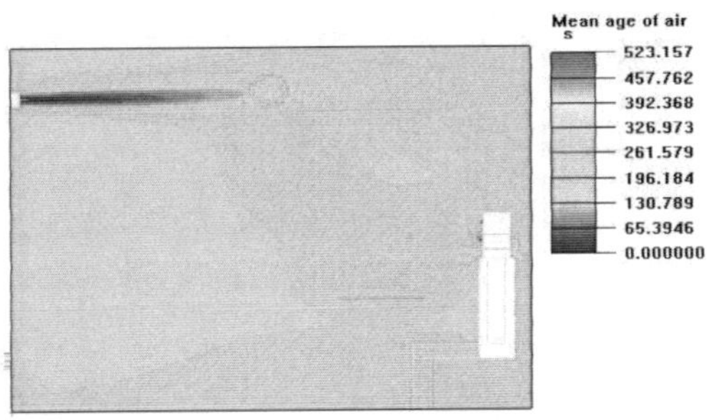

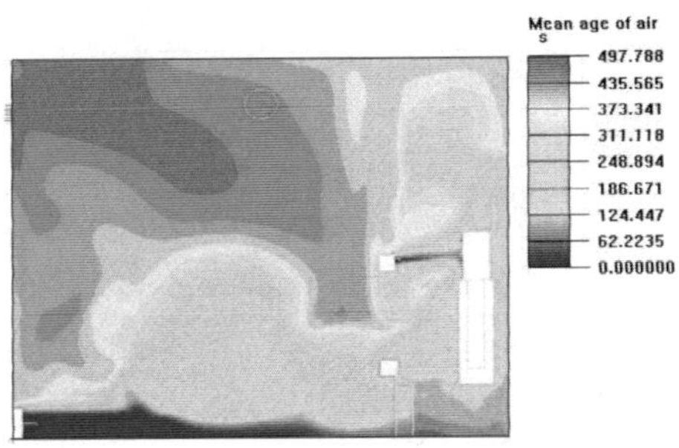

Fig. 10. Contour of mean age of air(°C) (z-plane) **Fig. 11.** Contour of mean age of air(z-plane)

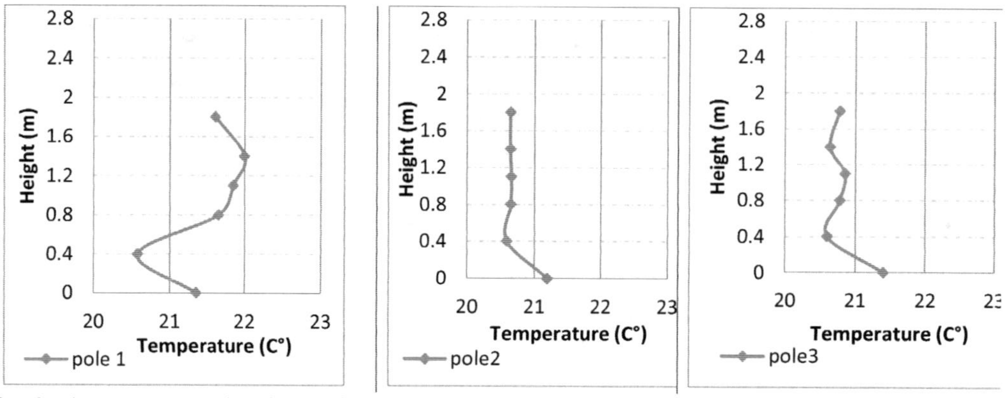

Fig.12. Air temperature distribution for case I

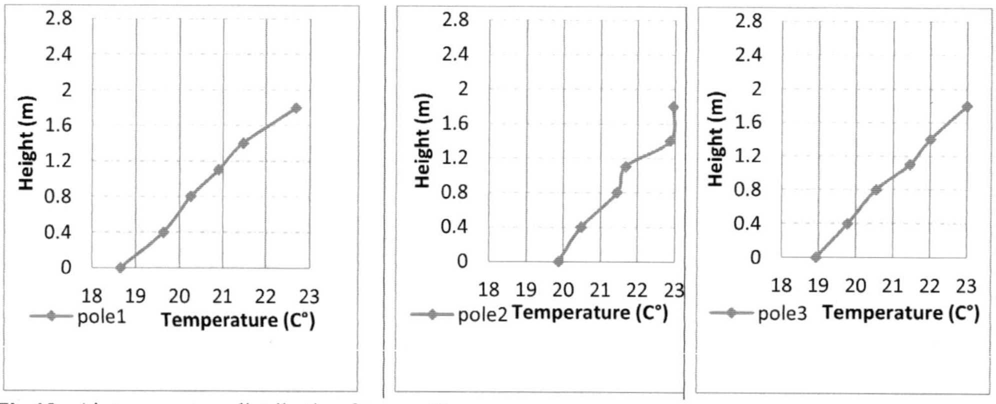

Fig.13. Air temperature distribution for case II

A Theoretical Study of Second Lyapunov Method on Stability of Zero Solution for a Delay Differential System

Anas A. Hijab[1]*, Muayyad M. Khalil[1, 2] *

[1,2] Department of Mathematics, College of Education for pure Sciences, Tikrit University.

[1]**Corresponding author:** anas_abass@tu.edu.iq

Abstract. This study examined the asymptotic stability of delay differential systems using Lyapunov's second method. It showed that the zero solutions are asymptotically stable when satisfying some necessary and sufficient conditions. The solution origin of the given system is asymptotically stable for small perturbations, which has no effect on equilibrium system if some conditions are held and it is asymptotically stable for large perturbations that moving system of its equilibrium then its comeback to it if other conditions are held. The proof of theorems in this article results by using and applying Gronwell-Bellman inequality which leads us to main result.

Keywords: Lyapunov function, Delay differential equations, Asymptotic, Stability.

1. Introduction

In the past ten years, the control community has intensively studied time delay systems TDS analysis and control in response to evolving theories of adapted control and an essential requirement of applications in information technology [1, 2, 3, 4]. In addition, they are known as systems with after effect that belong to the category of functional differential equations, which are infinite-dimensional contrasted with ordinary differential equations. Time-delays occur in various engineering systems including aircraft, medicine, chemical control, laser models and Internet biology [5, 6]. As the case with systems that are free of delay, the Lyapunov method is efficient for analyzing stability of TDS. For TDS, two main methods of Lyapunov exist: the Lyapunov-Krasovskii [7, 8, 9] and Lyapunov-Razumikhin. There are numerous authors dealt with stability of TDS as being studied in [10, 11, 12, 13, 14, 15]. In this article, the second Lyapunov method was employed to determine the asymptotic stability of a null solution of proposed system.

2. Preliminaries.

Definition (2.1): If F is a function from $T \times R^n$ to $T \times R^n$, where $T = [0, \infty)$ then F satisfies Lipschitz condition, if it satisfies the inequality in [2],

$$\|F(t, x_1, x_2, \dots, x_n) - F(t, y_1, y_2, \dots, y_n\| \le \sum_{k=1}^{n} l_k \|x_k - y_k\|,$$

Where $t \in [0, \infty)$.

Definition (2.2): Assume that $x(t) = \theta(t)$ is a solution for the system [5],

$$\dot{x}(t) = ax(t) \tag{1}$$

is defined for $t_0 \leq t$. Let $x(t) = \vartheta(t)$ be another solution for (1) with initial condition $x_0 = \vartheta(t_0)$, we say that the solution $x(t) = \theta(t)$ is stable wirh respect to Lyapunov's concept if there exist for any number $\epsilon > 0$ and $\delta > 0$ such that

$$\|\theta(t_0) - \vartheta(t_0)\| \leq \delta \Longrightarrow |\theta(t) - \vartheta(t)| \leq \epsilon$$

This means that if the solution $\theta(t)$ was closed from $\vartheta(t)$ at the initial condition, so it will be too for $t_0 \leq t$.

Definition (2.3): We say that the solution $x(t) = \theta(t)$ is asymptotically stable, if it is stable w.r.t Lyapunov's concept and $\lim_{t\to\infty} \|\theta(t) - \vartheta(t)\| = 0$ [6].

Now, we review some basics of Lyapunov's theorem of stability. If $v(x): D \to R$ is a function defined on $D \subseteq R^n$, then the cases are possible.

Case ***A***.

The solution origin of system (1) is asymptotically stable for small perturbations, which has no effect on equilibrium system if these conditions are held:

C1: $v(x) > 0$, for $x > 0$.

C2: $v(0) = 0$.

C3: $v(x)$ has partial derivatives.

C4: $\dot{v}(x) < 0$, for $x \neq 0$ and x lies at neighborhood of the origin.

Case ***B***.

The solution origin of system (1) is asymptotically stable for large perturbations that moving system of its equilibrium then it comeback to it if these conditions are held:

C1: $v(x) > 0$, for $x > 0$.

C2: $v(0) = 0$.

C3: $v(x)$ has a continuous partial derivative.

C4: $\dot{v}(x) < 0$, for $x \neq 0$ and x lies at neighborhood of the origin and satisfying the condition $v(x) \to 0$, when $\|x\| \to \infty$.

Remark (2.4): The function $v(x)$ is called Lyapunov's function, this function exist mean absolutely that the system is stable [10].

Definition (2.5): (Gronwall-Bellman inequality) Assuming $x(t)$ is a continuous function and [10]: $h(r) \geq 0$, if $x(t) \leq \omega(t) + \int_{t_0}^{t} h(r)x(r)dr, t \geq t_0$, where $\omega(t)$ is decreasing and integrable function then

$$x(t) \leq \omega(t)e^{\int_{t_0}^{t} h(r)dr}$$

IOP Conf. Series: Materials Science and Engineering 1095 (2021) 012002 doi:10.1088/1757-899X/1095/1/012002

3. Goal of Research

We will study the stability of this system of differential equation:

$$\dot{x}(t) = a(t)\,x(t) + b(t)\,x(t-\varphi). \tag{2}$$

In our work, we shall discuss the following two problems.

Problem $(\mathbf{3.1})$:

Studying stability of system (2) without delay that is $\varphi = 0$. So the system will become

$$\dot{x}(t) = G(t)x(t),$$

, where $G(t) = a(t) + b(t)$.

Problem $(\mathbf{3.2})$:

We are going to study the stability of system (2) as is.

4. Results and Discussion:

Suppose that the following system exists:

$$\dot{x}(t) = G(t)\,x(t). \tag{3}$$

We seeking the Lyapunov function whose quadratic formula is

$$v(x,t) = x^T H(t) x. \tag{4}$$

Differentiating (4) with respect to t we find

$$\dot{v}(\chi,t) = \dot{x}^T H(t)\chi + x^T \dot{H}(t)\chi + \chi^T H(t)\dot{\chi}$$

$$\Rightarrow \dot{v}(\chi,t) = \chi^T G^T H(t)\chi + \chi^T \dot{H}(t)\chi + \chi^T H(t)\dot{\chi}. \tag{5}$$

Substituting $\dot{x}$ into (5),

$$\Rightarrow \dot{v}(\chi,t) = \chi^T G^T H(t)\chi + \chi^T \dot{H}(t)\chi + \chi^T H(t)G(t)\chi. \tag{6}$$

Taking out the common factor $\chi\chi^T$,

$$\Rightarrow \dot{v}(\chi,t) = \chi^T \left(G^T H(t) + \dot{H}(t) + H(t)G(t)\right)\chi. \tag{7}$$

Assuming that

$$v(\chi,t) = -w(x,t), \tag{8}$$

where $J(t)$ is a square matrix of rank $n \times n$ and $w(t) = x^T J(t) x(t)$ may be determined from $G(t)$. From (7) and (8) we find that

$$x^T \left(G^T H(t) + \dot{H}(t) + H(t)G(t)\right) x = x^T J(t) x(t)$$

$$\Rightarrow G^T H(t) + \dot{H}(t) + H(t)G(t) = -J(t). \tag{9}$$

Multiplying each term of (9) by $e^{\int_0^t G^T(r)dr}$ and $e^{\int_0^t G(r)dr}$ from left and right respectively for $H(t)$

$$e^{\int_0^t G^T(r)dr} G^T H(t) e^{\int_0^t G(r)dr} + e^{\int_0^t G^T(r)dr} \dot{H}(t) e^{\int_0^t G(r)dr}$$

$$+ e^{\int_0^t G^T(r)dr} H(t)G(t) e^{\int_0^t G(r)dr} = -e^{\int_0^t G^T(r)dr} J(t) e^{\int_0^t G(r)dr}$$

$$\Rightarrow e^{\int_0^t G^T(r)dr} \dot{H}(t) e^{\int_0^t G(r)dr} = -e^{\int_0^t G^T(r)dr} J(t) e^{\int_0^t G(r)dr}$$

$$\Rightarrow \frac{d}{dt}\left(e^{\int_0^t G^T(r)dr} H(t) e^{\int_0^t G(r)dr}\right) = -e^{\int_0^t G^T(r)dr} J(t) e^{\int_0^t G(r)dr}.$$

Integrating for t,

$$e^{\int_0^t G^T(r)dr} H(t) e^{\int_0^t G(r)dr} = -\int_0^t e^{\int_0^\omega G^T(r)dr} J(\omega) e^{\int_0^\omega G(r)dr}\, d\omega$$

$$\Rightarrow H(t) = -e^{-\int_0^t G^T(r)dr}\left[\int_0^t e^{\int_0^\omega G^T(r)dr} J(\omega) e^{\int_0^\omega G(r)dr}\, d\omega\right] e^{-\int_0^t G(r)dr}$$

$$\Rightarrow H(t) = -\int_0^t e^{\int_0^\omega G^T(r)dr} J(\omega) e^{\int_0^\omega G(r)dr}\, d\omega.$$

Since each of $v(x,t)$ and $w(x,t)$ is strictly positive, there is a $\rho(t_0)$ so that

$$\rho(t_0)\cdot v(x,t) \le w(x,t) \Rightarrow \rho(t_0)\cdot v(x,t) \le -\dot{v}(x,t)$$

$$-\rho(t_0)\cdot v(x,t) \ge \dot{v}(x,t) \Rightarrow \frac{\dot{v}(x,t)}{v(x,t)} \le -\rho(t_0)$$

$$\ln v(x,y) \le -t\rho(t_0) + \ln v(0) \Rightarrow v(x,y) \le e^{-t\rho(t_0)} v(x_0,y_0).$$

But

$$v(x,t) \ge \alpha(t)\|x(t)\|^2 \tag{10}$$

$$\Rightarrow \|x(t)\|^2 \le \frac{1}{\alpha(t)} v(x,t) \tag{11}$$

$$\Rightarrow \|x(t)\|^2 \le \frac{1}{\alpha(t)} e^{-t\rho(t_0)} v(x_0,y_0), \forall\, t \in (0,\infty) \tag{12}$$

$$\Rightarrow \|x(t)\|^2 \le \frac{1}{\alpha(t_0)} e^{-t\rho(t_0)} v(x_0,y_0). \tag{13}$$

And we have

$$v(x,t) \le \beta(t)\|x(t)\|^2 \tag{14}$$

$$\Rightarrow v(x_0, t_0) \leq \beta(t_0)\|x(0)\|^2 \tag{15}$$

$$\Rightarrow \|x(t)\|^2 \leq \frac{1}{\alpha(t_0)} e^{-t\rho(t_0)} v(x_0, y_0) \leq \frac{\beta(t_0)}{\alpha(t_0)} e^{-t\rho(t_0)} \|x(0)\|^2 \tag{16}$$

$$\Rightarrow \|x(t)\|^2 \leq \frac{\beta(t_0)}{\alpha(t_0)} e^{-t\rho(t_0)} \|x(0)\|^2. \tag{17}$$

So, the last inequality gives us the necessary condition for system's stability

$$\lim_{t\to\infty} \|x(t)\|^2 = 0.$$

Hence, (3) is asymptotically stable.

We now begin our study of the stability of system (2)

$$\dot{x}(t) = a(t) \cdot x(t) + b(t) \cdot x(t - \varphi). \tag{18}$$

We will take a special case of Gronwall-Bellman inequality

$$x(t) = e^{\int_0^t a(r)dr} u(t). \tag{19}$$

Substituting (19) into (18), we see that

$$\dot{x}(t) = a(t) e^{\int_0^t a(r)dr} u(t) + e^{\int_0^t a(r)dr} \frac{u(t)}{dt}$$

$$\Rightarrow a(t) \cdot x(t) + b(t) \cdot x(t - \varphi) = a(t) e^{\int_0^t a(r)dr} u(t) + e^{\int_0^t a(r)dr} \frac{u(t)}{dt}$$

$$\Rightarrow a(t) e^{\int_0^t a(r)dr} u(t) + b(t) e^{\int_0^{t-\omega} a(r)dr} u(t - \omega) = a(t) e^{\int_0^t a(r)dr} u(t) + e^{\int_0^t a(r)dr} \frac{u(t)}{dt}$$

$$\Rightarrow \frac{u(t)}{dt} = b(t) e^{\int_0^{t-\omega} a(r)dr - \int_0^t a(r)dr} u(t - \omega)$$

$$\Rightarrow \frac{u(t)}{dt} = b(t) e^{\int_t^{t-\omega} a(r)dr} u(t - \omega).$$

Moreover, if taking a quadratic Lyapunov function $v = u^T H(t) u$, notice that, its hold the inequality $\alpha\|u\|^2 \leq v(x) \leq \beta\|u\|^2$.

Differentiating, we see that

$$\dot{v} = \dot{u}^T H(t) u + u^T \dot{H}(t) u + u^T H(t) \dot{u}$$

$$\Rightarrow \dot{v} = u^T(t - \omega) e^{\int_t^{t-\omega} a^T(r)dr} b^T H(t) u(t) + u^T \dot{H}(t) u + u^T H(t) b(t) e^{\int_t^{t-\omega} a(r)dr} u(t - \omega).$$

Integrating from zero to t,

$$v(x) = v\big(x(0)\big) + \int_0^t u^T(r-\omega)e^{\int_r^{r-\omega} a^T(z)dz} b^T(r)H(r)u(r) + u^T(r)\dot{H}(r)u(r)$$

$$+u^T(r)H(r)b(r)e^{\int_r^{r-\omega} a(z)dz} u(r-\omega)dr.$$

Using the inequality

$$\alpha\|u\|^2 \le v(t) \le \beta\|u\|^2.$$

We see that

$$\Rightarrow v(x) \le v\big(x(0)\big) + \frac{1}{\alpha}\left[\int_o^t \sqrt{v^T(r-\omega)}\, e^{\int_r^{r-\omega} a^T(z)dz} b^T(r)H(r)\sqrt{v(r)} + \sqrt{v^T(r)}\dot{H}(r)\sqrt{v(r)} + \sqrt{v(r)}H(r)b(r)e^{\int_r^{r-\omega} a(z)dz}\sqrt{r-\omega}\right] dr.$$

By taking the non-decreasing norm $\|v(t)\|_I = \sup\|v(r)\|$, where $-\omega \le r \le \omega$, it is observed that

$$\Rightarrow v(x) \le v\big(x(0)\big) + M + \frac{1}{\alpha}\int_0^t \Big\{\|v(r)\|e^{\int_r^{r-\omega}\|a(z)\|dz\|b(r)\|\|H(r)\|} + \|v(r)\|\|\dot{H}(r)\| + \|v(r)\|\|H(r)\|\|b(r)\|e^{e^{\int_r^{r-\omega}\|a(z)\|dz}}\Big\}dr$$

$$\Rightarrow v(x) \le \big(v\big(x(0)\big) + M\big) + \frac{1}{\alpha}\int_0^t \Big\{2\|b(r)\|\|H(r)\| + e^{e^{\int_r^{r-\omega}\|a(z)\|dz}} + \|\dot{H}(r)\|\Big\}\|v(r)\|dr.$$

Applying Gronwell-Bellman inequality, we see that

$$\Rightarrow v(x) \le \big(v\big(x(0)\big) + M\big)e^{\frac{1}{\alpha}\int_0^t\Big\{2\|b(r)\|\|H(r)\|+e^{e^{\int_r^{r-\omega}\|a(z)\|dz}}+\|\dot{H}(r)\|\Big\}dr}.$$

By inequality, we have that

$$\alpha\|u\|^2 \le v(t) \le \beta\|u\|^2$$

$$\Rightarrow \alpha\|x\|^2 \le (\beta\|x\|^2 + M)e^{\frac{1}{\alpha}\int_0^t\Big\{2\|b(r)\|\|H(r)\|+e^{e^{\int_r^{r-\omega}\|a(z)\|dz}}+\|\dot{H}(r)\|\Big\}dr}$$

$$\Rightarrow \|x\|^2 \le \frac{1}{\alpha}(\beta\|x\|^2 + M)e^{\frac{1}{\alpha}\int_0^t\Big\{2\|b(r)\|\|H(r)\|+e^{e^{\int_r^{r-\omega}\|a(z)\|dz}}+\|\dot{H}(r)\|\Big\}dr}.$$

Hence, this is a condition for the system stability. Finally, it is observed that the criterion of stability is satisfied when the square of the norm does not exceed

$$\frac{1}{\alpha}(\beta\|x\|^2+M)e^{\frac{1}{\alpha}\int_0^t\left\{2\|b(r)\|\|H(r)\|+e^{e^{\int_r^{r-\omega}\|a(z)\|dz}}+\|\dot{H}(r)\|\right\}dr}.$$

Moreover, the criterion is satisfied when

$$\lim_{t\to\infty}\frac{1}{\alpha}\int_0^t\left\{2\|b(r)\|\|H(r)\|+e^{e^{\int_r^{r-\omega}\|a(z)\|dz}}+\|\dot{H}(r)\|\right\}dt<0.$$

We note that condition of stability depends on delay ω and $a(t), b(t)$.

5. Conclusions:

We proved in this article the asymptotic stability of the null solution of proposed system (2), that is if the trivial solution of the system, $\dot{x}(t)=(a(t)+b(t))x(t)$ is, asymptotically stable and formula, $\|x\|^2\leq\frac{1}{\alpha}(\beta\|x\|^2+M)e^{\frac{1}{\alpha}\int_0^t\left\{2\|b(r)\|\|H(r)\|+e^{e^{\int_r^{r-\omega}\|a(z)\|dz}}+\|\dot{H}(r)\|\right\}dr}$ holds, then the zero solution for (2) is asymptotically stable too, when $M=\underbrace{\sup v(x(t))}_{-\omega\leq t\leq 0}$ and $v(x(t))$ is the Lyapunov function and α,β are eigenvalues for the matrix G. For the future works, we suggest to authors to study the asymptotic stability of solutions of stochastic differential delay equations by using the technique of Lyapunov-Krasovskii functional. So, may someone can study Lyapunov's second method in problems of the stability of solutions of systems with impulse effect.

References

[1] Gopalsamy K. Stability and oscillations in delay differential equations of population dynamics. Springer Science & Business Media; Vol. **74** *Mathematics and Applications*.

[2] Richard JP. Time-delay systems: an overview of some recent advances and open problems. *automatica*. 2003 Oct 1;**39**(10):1667-94.

[3] Li X, De Souza CE. Delay-dependent robust stability and stabilization of uncertain linear delay systems: a linear matrix inequality approach. *IEEE Transactions on Automatic Control*. 1997 Aug;**42**(8):1144-8.

[4] Fridman E, Shaked U. Delay-dependent stability and H∞ control: constant and time-varying delays. *International journal of control*. 2003 Jan 1;**76**(1):48-60.

[5] Hale J.K.,S.M.VerduynLunel. *Introduction to Functional Differential Equations*. Springer-Verlag,New-York. 1993.

[6] Kolmanovskii V, Myshkis A. *Applied theory of functional differential equations*. Springer Science & Business Media; 2012 Dec 6.

[7] Krasovskii NN. *Stability of motion*. Stanford University Press. 1963.

[8] Fridman E. New Lyapunov–Krasovskii functionals for stability of linear retarded and neutral type systems. *Systems & control letters*. 2001 Jul 23;**43**(4):309-19.

[9] Burton T. A. . *Stability and periodic solutions of ordinary and functional differential equations*. Mathematics in Science and Engineering. 178. Academic Press. Inc. Orlando. FL. 1985.

[10] Burton TA, Hering RH. Liapunov theory for functional differential equations. *The Rocky Mountain Journal of Mathematics*. 1994 Jan 1;**24**(1):3-17.

[11] Oudjedi L, Beldjerd D, Remili M. On the stability of solutions for non-autonomous delay differential equations of third-order. *Differential Equations and Control Processes*. 2014 Mar **1**;1(2014):22-34.

[12] Burton T, Hatvani L. Stability theorems for nonautonomous functional differential equations by Liapunov functionals. Tohoku Mathematical Journal, Second Series. 1989;**41**(1):65-104.

[13] Makay G. An example on the asymptotic stability for functional differential equations. *Nonlinear Analysis: Theory, Methods & Applications*. 1994 Aug 1;**23**(3):365-8.

[14] Omeike M. O. . New results on the stability of solution of some non-autonomous delay differential equations of the third order. *Differential Equations and Control Processes.* 2010; **1**, 18–29.

[15] Sadek AI. On the stability of solutions of some non-autonomous delay differential equations of the third order. Asymptotic Analysis. 2005 Jan 1;**43**(1, 2):1-7.

IOP Conf. Series: Materials Science and Engineering 1095 (2021) 012003 doi:10.1088/1757-899X/1095/1/012003

An Experimental study of some Physical properties on Gallium Antimonide Nanocrystals

Saad Majeed[1], Bilal K. Al-Rawi[2]

Department of Physics, College of Education for Pure Science, University of Anbar, Anbar; Iraq.

Corresponding author: [1]saad.majeed70@yahoo.com; [2]sc.bilal_alrawi@uoanbar.edu.iq

Abstract. This study includes preparing Nano-thin films of gallium antimonide (GaSb) and study the physical properties of this films. These films are prepared by using a vacuum evaporation technique with pressure (10^{-6} Torr). They were deposited on glass substrates. After the preparation process, some of its optical and structural properties, such as (energy gap, optical absorption, AFM, SEM and XRD) were studied. By discussing the results, it was found that the prepared GaSb thin films have worthwhile structural and optical properties. These properties can be used in the preparation and manufacture of infrared devices, laser detectors and photoelectric applications. The distinguished obtained characteristics are of interest to a large number of researchers within the fields of semiconducting materials. The potential applications that could employ GaSb Nano-thin films in their work are infrared devices and wide range applications of detectors and photo electronics. For example, Li-Fi technology, which relies on visible light as a means of transmitting data instead of radio frequencies, in addition to it does not generate any radiation as in Wi-Fi, as it ultimately depends on visible light.

Keywords: GaSb, nanomaterials, structure properties, nanocrystals.

1. Introduction

Gallium Antimonide (GaSb) is a crystalline semiconductor compound of gallium and antimony of the III-V family [1]. The crystal structure of Gallium Antimonide (GaSb) is Zinc Belende [2]. Its Lattice constant about 6.0959 Å and the density (gm cm^{-3}) is 5.6137 [3]. The energy gap is 0.726 eV in the bulk structure [4,5]. GaSb is a direct gap semiconductor (0.72 eV) capable of being doped either *p* or *n* type with good mobilities and it has significant electro-optical potential in the near IR range [5,6]. As a substrate, or active layer, GaSb can be employed in conjunction with many semiconductors, such as (AlGa)Sb or In(AsSb). It has interesting heterojunction potential for detectors and lasers and quantum well structures [6]. Figure (1) shows the GaSb crystal structure.

The semiconductor within the infrared III-V groups is one of the most candidate materials of using in lasers and detectors [7,8], because it possesses transactions with high quantum efficiency. Its melting point is between 705 and 712 C°, so it is easy to prepare and handle it physically [9,10]. Absorption capacity and high transport traffic are the most important features of III-V groups and it possesses wide-range adjustable loopholes [11]. Among these materials is gallium-based alloys and compounds. GaSb is the most important one, as it provides a wide range of electronic energy gaps. This wide range makes a very high electron movement along with electronic barriers. This is to produce a group of devices working speedily. The devices work depends on GaAs, InGaSb and InP, in addition, they have lowest energy consumption [12].The superstructure of this material is the quantum well which has a great interest. In addition, a variety of advanced optical electronic devices, such as laser diodes, detectors, and transistors have been achieved by using these materials from (GaSb) [9]. Consequently, this study will present the preparation of three different thicknesses of GaSb membranes. Also, it presents some structural and optical properties of the prepared membranes.

2. Experimental Part

In the experimental part of this study, there are two methods in order to prepare the Gallium Antimonide Nano-thin films.

*2.1. Method of preparing gla*ss slides.

For the purpose of deposition, the glass slides are used by (2.5×2.5×0.1 cm). These slides have washed by cleaning powder and normal water to remove dust. After that, the slides are immersed in chemical vessel contains a distilled water. Then, the chemical vessel is putted in ultrasonic for 10 minutes. The following step, the slides are immersed in Ethanol solution for 10 minutes in ultrasonic in order to remove fatty residues which remain on the slides surfaces. Subsequently, these slides are wiped by cleaning paper carefully to be ready for deposition. Consequently, they have cut into smaller slides (2.5×2.5) dimensions per cm. These slides are individually weighted using the sensor sensitive scale (Mettler type) which has a 10-4 sensitivity. The mass required for each of the three deposition processes was (m = 0.0705 g).

2.2. Method of preparing Gallium Antimonide Nano-thin films.

The deposition process was done under pressure 10-6 Torr where the adopted height in the first model was (20 cm), the second height (18 cm) and the third height (16 cm). Through the three aforementioned heights three films of different thicknesses were obtained.

3. Results and Discussions

Figure (2) shows the values of the optical energy gap (Eg) for (GaSb) thin films prepared by vacuum thermal evaporation resulting from the graph relationship between ($\alpha * h\nu$ and). This indicates the good preparation conditions, as well as the thermal annealing process that gave the thin films distinctive homogeneity. For the first film, with an approximate thickness (60 nm) the energy gap (1.45 eV), the second film with an approximate thickness (78 nm) the energy gap (1.68 eV), and the third film, with an approximate thickness (85 nm), its energy gap (1.87 eV). It is worth noting that the energy gap increased from the usual value in the sensed condition, which is around (0.726 eV) [4,5]. As for the simple differences between the three values due to the fact that the theoretical method is of ideal standards. As for the practical method, on the contrary, it is affected by the conditions of preparation and the processes associated with it.

The results of the X-ray diffraction spectrum as shown in Figure (3) showed that the first thin film has a random crystal structure (Amorphous). Other films, the nature of the crystal structure has polycrystalline and Hexagonal, Tetragonal, Orthorhombic, and Tetragonal structures respectively. It was in the preferred direction (003). The crystalline levels of the prepared thin film structures were (003), (200), (200), (121), respectively. They were found to be consistent with (JCPDS) cards for (Ga, Sb, and GaSb). The first crystalline level (003) belong to Antimone and, which complies with JCPDS card (24, 1968-1982, 38), the two levels (200) belong to Gallium (Ga) and are identical to the (JCPDS) cards (54, 2768-2770, 1971) and (18, 1978-2009, 80). The level (121) refers to the (GaSb) compound, which Complies with JCPDS card (36, 1498-1499, 1965). Decreasing the height between the glass base platform and the GaSb crucible leads to a significant increase in the height of the X-ray diffraction peaks. This is also due to the difference in evaporation temperatures for (Ga) and (Sb). Two peaks appeared for (Ga) and a peak for (Sb). This means that X-ray suffered greater diffraction at (Ga) than (Sb). The clear and sharp appearance of peaks in the third thin film is caused by an increasing in the density of atoms on the surface of that thin film, thus may cause greater (x-ray) diffraction compared to other films. The Debye_Scherer equation was used to calculate the crystal size of the prepared films as shown in Table (1); and the average crystal size was calculated in the order of (13 nm).

Figure (4) shows that the surface of the first thin film is composed of nanoparticles with spherical, semi-oval shapes. It has a homogeneous distribution, high density on the surface of the nanoparticle and the granular boundaries as it appears less than the radius of the crystalline grain with respect to the first thin film. The diameter of the grains appears in the microscopic images as an average for thin film grains in general and for each sample. In the corresponding image, which is within the macrometer scale (μm), the distribution of particles is homogeneous and uniform largely. There are some pits or clumps which may be produced due to the conditions of preparation or storage. As for the second thin film, we notice an increasing in the granular diameter with an increasing in the intergranular boundaries of the granules forming the surface. The distribution remains homogeneous and uniform during the sedimentation process in an identical manner to the surface of the first thin film. The third thin film showed a significant increase in the granular diameter as well as the grainy boundaries which compared to the first and second thin films with a homogeneous distribution of the granules on the surface. The third thin film pictures also showed crystal masses of relatively irregular shapes as well as with varying diameters between the nanometer and the macrometer scale. This matter can be attributed to the crystalline growth of the thin film surface in the third sample has become vertical.

Figure (5) shows the microscope images that the films surface possesses homogeneity, size, and granular shape. This depends on the experiment variable, which is the thickness resulting from the change in the height values between the sedimentation pot (the powder) and the platform on which the glass bases are fixed in the evaporating chamber. Table (2) shows the average grain size (nm), Root mean square (nm), and roughness average (nm) in the AFM scan. The average grain size ranges from (69.31 nm) to (87.13 nm) and depends on the conditions of sedimentation. As for the root mean square values, they range between (6.13nm) and (7.79 nm). The AFM images show the first thin film (N1) with a homogeneous surface with a grain size (87.13 nm) and a roughness rate (5.31 nm). The second thin film (N2) shows a decrease in the average grain size (85.14 nm) with an increase in the roughness average (6.9 nm). Consequently, the third thin film (N3) shows a decrease in the average grain size (69.31 nm) compared to the second thin film. So that, the roughness average shows a decrease to (6.55 nm) as the reason of changing in the roughness of the thin films was due to the change in the growth of the thin films. Associated with the speed of the deposition process, in which the third thin film will be higher compared to the first and second thin film. This is due to the decrease in the height between the fineness of the preparation and the glass bases. These properties and variation in roughness can be invested in industrial applications such as optical devices and photoelectric applications.

Figure (6) shows that the absorbance values are almost constant in the infrared region, and then they gradually increase in the visible region. By reaching the wavelengths confined between (300 nm) and (500 nm) and reach maximum values at the wavelength (300 nm). The figure shows that the absorbance behavior in the three thin films is almost identical, and it appears that the first thin film has the highest absorbency. It also notice that the third thin film has the lowest absorption, due to the difference in the thickness of the prepared nanofilms. It appears that the thin film which is the largest value of the thickness has the largest value of absorbance compared to the rest of the films. The surface of the larger thin film will suffer greater attenuation if the light spectrum falls when it interacts with the surface grains, as well as when it penetrates the thin film to a certain depth. and as such, the absorbance of the thin film will be high in relation to thicker thin film, and greater absorption.

4. Conclusions

The results of the study showed that the optical energy gap increased from the value in the palpable condition (0.726 eV), as it was the highest in the third thin film (1.87 eV). As for the optical absorbance spectrum, it showed that the first thin film had the highest absorbance. Also for the (AFM) images, they showed a characteristic homogeneity for all surfaces of the prepared thin films, and the results of (AFM) indicated that the average grain size decreased (from 87.13 nm to 69.31 nm) with an increase in the rate of Roughness (from 5.31 nm to 6.55 nm). The X-ray examinations showed the characteristic trend of the crystal levels (003) as well as the average crystal size (13 nm) in addition to the geometric composition of the crystal lattice for each peak of the X-ray diffraction peaks.

References

[1] R. A. Laudise 1983 *J. Cryst. Growth* **65** 3-23.

[2] National Bureau of Standards (U.S.) 1956 Circ. 539 **6** 12.

[3] V. Swaminathan and A. T. Macrander, 1991 Materials spects of GaAs and InP.based structures *Prentice–Hall Englewood Cliffs NJ* **Chap.1**.

[4] A. Licht, N. Pfiester, D. DeMeo, J. Chivers, and T. E. Vandervelde 2019 *MRS Adv* **4** 2271.

[5] M. Levinštejn, S. Rumyantsev, and M. Shur (ed.) 1996 *Handbook Series on Semiconductor Parameters World Scientific* Singapore.

[6] H. Xie, J. Piao, J. Katz, and W. I. Wang *J.* 1991 *Appl. Phys.* **70** 3152.

[7] G. Motosugi and T. Kagawa, Jpn 1980 *J. Appl. Phys* **19** 2303.

[8] M. B. Z. Morosini, J. L. Herrera-Perez, M. S. S. Loural, A. A. G. Von Zuben, A. C. F. da Silveira, and N. B. Patel 1993 *IEEE J. Quantum Electron QE* **29** 2103.

[9] O. Hildebrand, W. Kuebart, K. W. Benz, and M. H. Pilkuhn 1981 *IEEE J. Quantum Electron. QE* **17**, 284; see also O. Hildebrand, W. Kuebart, and M. H. Pilkuhn 1980 *Appl. Phys. Lett.* **37**, 801.

[10] Constitution of Binary Alloys 1958 edited by *M. Hansen -McGraw–Hill* New York, p. 755.

[11] L. Esaki 1981 *J. Cryst. Growth* **52** 227.

[12] G. Muller 1990 *J. Cryst. Growth* **99** 1242; also see the NETSCAPE homepage: http://www.space.ca/cfzf/bkgnd.html

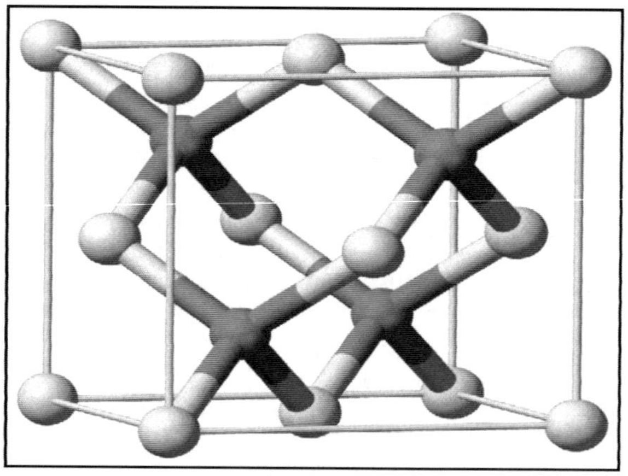

Figure (1): Shows the GaSb Crystal Structure.

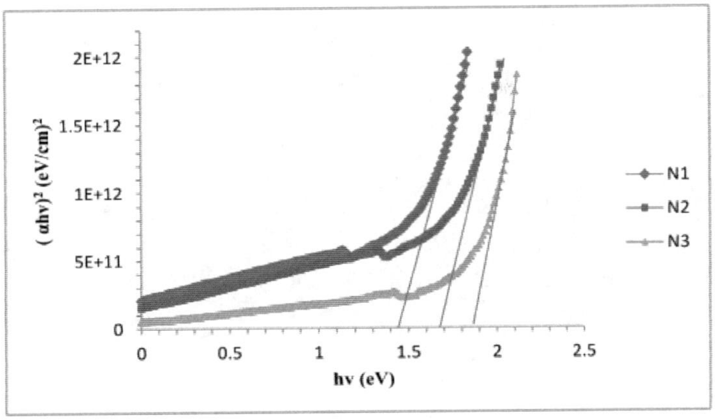

Figure (2): Shows the optical energy gap of (GaSb) thin films.

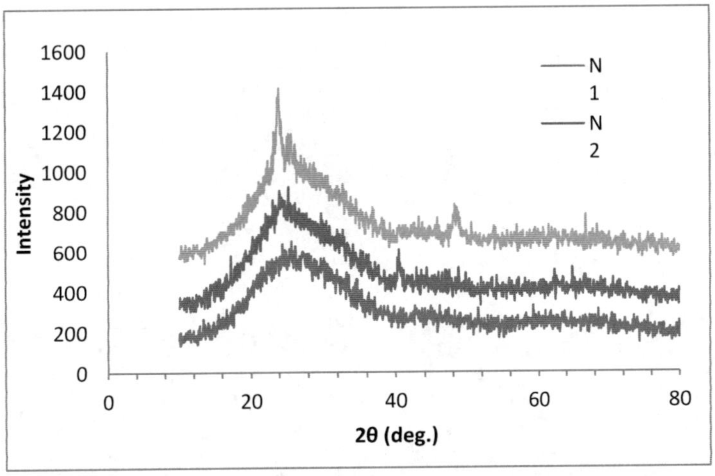

Figure (3): Shows the XRD patterns of (GaSb) Nano-thin films.

IOP Conf. Series: Materials Science and Engineering 1095 (2021) 012003 doi:10.1088/1757-899X/1095/1/012003

N1

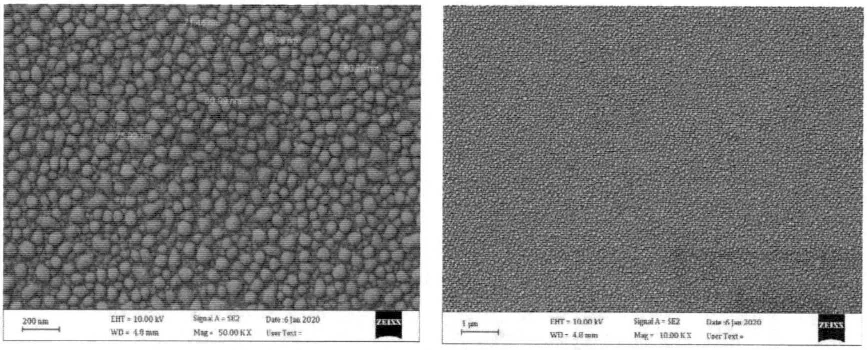

N2

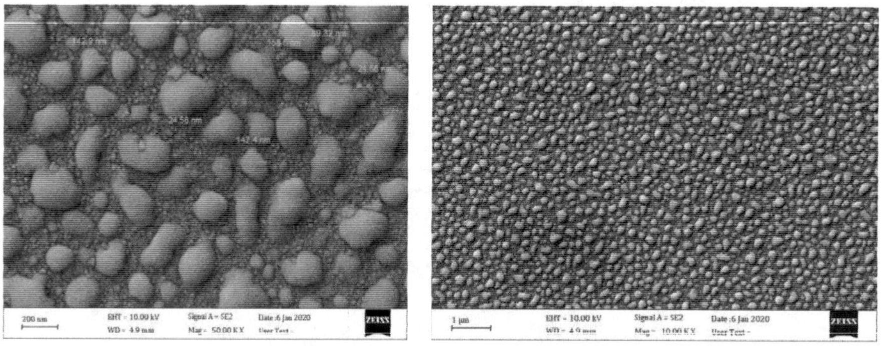

N3

Figure (4): Illustrates the images of the scanning electron microscope of GaSb Nano-thin films.

Table (1): XRD measurements of (GaSb) nanoparticles.

Sample No.	2θ	hkl	FWHM	d_{hkl} Exp.(Å)	d_{hkl} Std.(Å)	Gr.S (nm)	System
1	Amorphous						
2	24.19	003	3.936	3.678	3.523	2.16	Hexagonal
	40.40	200	0.393	2.232	2.229	22.51	Tetragonal
3	23.83	200	0.492	3.732	3.886	17.25	Orthorhombic
	48.42	121	0.787	1.879	1.912	11.57	Tetragonal

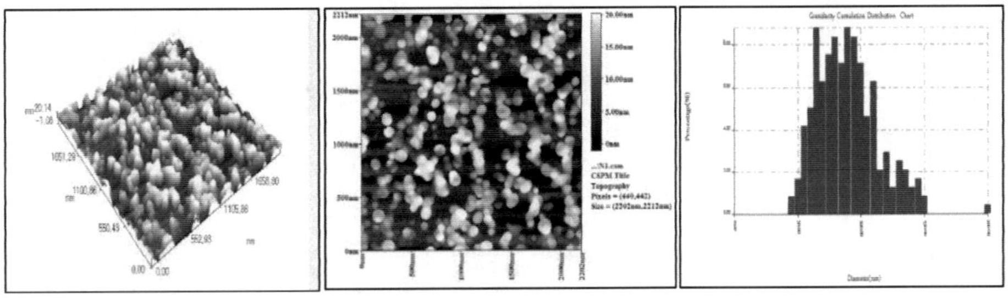

N1

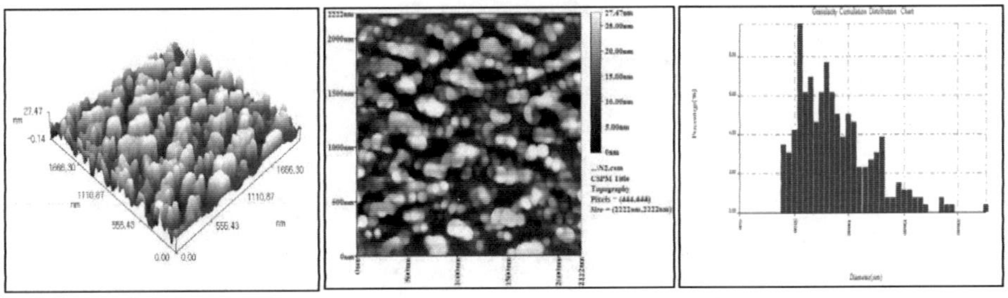

N2

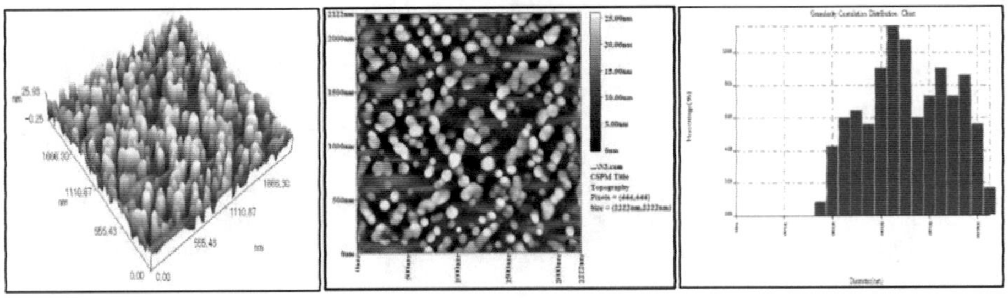

N3

Figure (5): Shows AFM images of GaSb thin films.

IOP Conf. Series: Materials Science and Engineering 1095 (2021) 012003 doi:10.1088/1757-899X/1095/1/012003

Table (2): AFM results of GaSb thin films.

Sample No.	Roughness Average (nm)	Root mean square (nm)	Average grain size (nm)
1	5.31	6.13	87.13
2	6.9	7.97	85.14
3	6.55	7.56	69.31

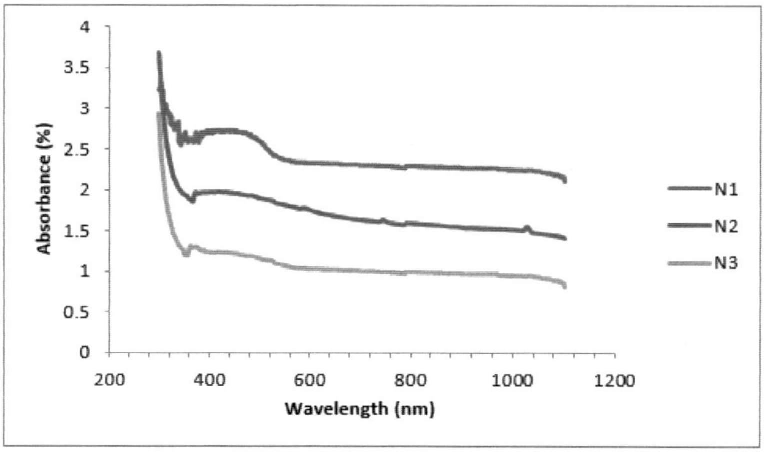

Figure (6): Shows the absorbance as a function of wavelength of GaSb Nano-thin films.

IOP Conf. Series: Materials Science and Engineering 1095 (2021) 012004 doi:10.1088/1757-899X/1095/1/012004

Determined the Effective Temperature and Gravity Acceleration at the Atmosphere of (HD165908) Star

Omar Ahmed Alhawi[1] and Firas Saleem Mazyed[2]

AMA International University, BAHRAIN,

Department of Physics, Wasit University, Iraq

dromaralhawi@gmail.com , fsaleem@uowasit.edu.iq

Abstract: The study of the chemical composition of stars in our Galaxy showed that it changes in a non-random way from a star to a star: older stars and halo stars have a metal deficit compared to young stars and disc stars. However, this is not always the case. Sometimes old stars have a solar chemical composition, and young stars show a deficit of metals. It is believed that the formed stars has a chemical composition of the interstellar substance from which it was formed; therefore, the initial content of elements in its atmosphere depends on the time and place of birth. Within the framework of this work, echelle spectra of star 99Her of spectral classes A-F has been proceed, obtained with a 1.5-meter telescope PTT-150 (Russian Turkish Telescope-150). As a result of processing the spectra, the equivalent widths of unbend and weakly blended absorption lines were measured, 350 lines this star. The measured equivalent line widths can be used for other studies of the atmospheres of the A-F stars composition. Stars with temperatures of 8000-9000 K show a significant excess of elements heavier than iron and some deficiency of light elements. These results, apparently, are a consequence of the effects of light pressure and gravitational diffusion of atoms and ions in atmospheres of calm and radiative atmospheres of A-stars.

Keywords: chemically peculiar, abundances-stars, chemical compositions.

1. Introduction

The study of the chemical composition of stars is important both in terms of determining the fundamental characteristics of stellar atmospheres, and in studying the evolution of stars and matter of the Galaxy.

The number and relative content of chemical elements in stars and interstellar medium depend on many different processes. For example, the history of star formation, the exchange of mass with the surrounding galaxy environment, the form of the initial mass function (IMF) and how the stars of different masses evolve. The study of the chemical composition of stellar atmospheres is, in fact, the main and most valuable source of information on the evolution of matter in the interior of stars, in the interstellar medium, and in the volume of the Galaxy as a whole. [1]

Theoretical models of the chemical evolution of the Galaxy try to explain the observed ratios of the contents of elements relative to iron, as well as the radial gradient of the contents of elements along the radius of the Galaxy. In addition to ours, other galaxies were formed from gas that did not contain heavy elements. During the Big Bang, only the lightest elements were synthesized, such as helium, lithium, and beryllium.[2]

The synthesis of heavier nuclei occurred already inside the stars. Some of these nuclei were formed at fairly quiet stages of the evolution of the star, and they fell into the interstellar medium through intense mass loss (in the Asymptotic Branch of the Giants, for example). Other elements were synthesized and thrown into the surrounding space during supernova explosions. Thus, the interstellar medium was enriched with relatively heavy elements. The enriched substance served as the starting material for the formation of the next generation of stars, which, in turn, also synthesized metals and threw them into the interstellar medium. That is, there was a cycle of matter in the Galaxy, during which the gas periodically fell into the next star, where he changed his chemical composition. The process depended on the rate of star formation in various regions of the Galaxy, as well as on the interstellar gas density. [3]

2. A- Stars and Early F-Stars

Stars of spectral class A and early subclasses F have an effective temperature in the range of 7000 - 10000 K. Considering data on the chemical composition, several types can be distinguished here, including chemical peculiarities [3]:

1. "Normal" stars of class A.
2. Magnetic Ap stars.
3. Am-stars.

Let us consider each group separately.

"Normal" stars

They are stars with solar chemical composition. For various reasons, such stars are difficult to indicate. For a long time, some stars were used as standards in the analysis of the chemical composition of other A-stars. Often, those stars that were believed to have a normal chemical composition later turned out to be of type Am. Quite often the only exact criterion for belonging to this type was the excess of heavy elements. Am stars do not differ from normal stars in the sense that they usually lack a magnetic field and variability of brightness. The content of relatively light metals is often close to solar. But only with careful analysis of the spectral lines of heavier metals, it can be established that the star is of type Am stars. A typical example is the Sirius star, previously classified as normal star.

You can also specify Vega. Studies have shown that its composition is different from solar (it has a general deficit of metals). Vega refers to the class of luminosity V and, possibly, to stars of the type Lambda Boo, surrounded by a dust envelope, masking the real chemical composition of the star. Reduced metal in the spectral class A can be observed not only in dwarfs, but also in giants. For example, HR7338 (αDra), belonging to the class A0 III. In A-supergiants, the content of heavy elements does not show large differences from solar elements. Perhaps small deviations from the normal chemical composition among supergiants in the vicinity of the Sun reflect the corresponding inhomogeneities in the chemical composition of the interstellar medium. It is also likely that the parameters Teff and log (g) could be inaccurately determined. [4]

In a number of cases, the metal deficit may be due to the unaccounted factor of duality. An example is the double system πSgr, consisting of two giants of class F, which, apparently, have a solar chemical

composition. However, if we study πSgr without regard for duality, a fictitious deficit of many elements is observed.
Attempts were made to find stars belonging to the early subclasses A, having narrow spectral lines, and which would have a solar chemical composition. But all the stars studied had any deviations from the normal composition, hence the conclusion was made that near the subclass A0, normal stars are extremely rare.
There is an assumption that normal A-stars belonging to the main sequence, with slow rotation (v <100 km / s), mostly evolve in Am. Therefore, it is unlikely to find a star with a solar chemical composition among them. However, at a rotation speed exceeding 100 km / s due to meridional circulation (stirring) caused by rotation, chemical anomalies can't be retained in the atmosphere. It is logical to assume that among fast-rotating A-stars, many should have a normal chemical composition. Unfortunately, it is rather difficult to investigate such objects, since the lines in their spectra are very wide due to rotation. It is especially difficult to determine the content of heavy and rare-earth elements, since their lines are weak and few. The most favorable option, when the speed of rotation is large, but the star we see from the pole. However, this situation is extremely rare. [5]

Am-stars.
They are stars with a high content of heavy elements. This group, quite numerous, is sometimes called "metallic" (metallic-line stars) stars. They have an effective temperature ranging from 7300 to 10000K. And log (g) from 3.5 to 4.5 (belong to the classes of luminosity V-IV). All Am-stars are near the main sequence, that is, when leaving it, the chemical anomalies anomalies disappear.
The chemical composition of elements of the iron group is usually close to solar, or slightly elevated. In general, there is a reduced or normal content of light elements and an increasing excess of heavy metals as the ordinal number increases. At a speed of rotation at the equator greater than 100 km / s (this is the limiting speed), chemical anomalies disappear.
There are a high number of cases Am stars are double, since the "visible" chemical composition of a binary system, determined without regard for duality, is far from the real chemical composition of the components. Thus, in order to identify real deviations, it is necessary to study the composition of each component of the binary star separately.
For Am stars, there is no noticeable magnetic field, and besides, they have no noticeable variations in the spectral lines that could tell us about the presence of inhomogeneities in the chemical composition on the surface of stars.

Ap-stars.
These stars detected magnetic fields reaching 30 kg. At the same time, Ap-stars show significant anomalies in the chemical composition.
Ap-stars are usually divided into two large subgroups-SrCrEu and Si. For some objects a refined classification is applied, for example SrCr, SrEu, CrEu. The belonging of the star to the SrCrEu subgroup does not at all mean that all three elements in it differ in anomalous content from other lines in terms of their anomalous contents.
The SrCrEu subgroup belongs to the effective temperature range from 7500 to 10000K. The subgroup Si corresponds to higher temperatures, from 9000 to 16000K. The values of log g correspond to GP stars (3.8 - 4.2).

The chemical composition of elements of the iron group is usually close to solar, or elevated.
Super-heavy metal lines (Z> 72), such as W, Pb, U, are observed in the spectra of Ap stars. But they are few and often blended. In the Si subgroup, because of the high temperatures, the lines of super heavy elements can be observed only with strong excesses of these elements.
The magnetic fields of Ap-stars often show periodic changes. With the same period, shine and color change. Also, the lines of some elements in the spectrum change. Magnetic and spectral variability can be explained if we assume the model of an inclined rotator. In this model, the axis of the dipole magnetic field, the axis of rotation of the star, the line of sight do not coincide with each other. Rotation leads to periodic changes in the spectral lines, if it is assumed that on the surface of the star the corresponding elements are concentrated in spots. When the star is visible from the pole of rotation, the magnetic and spectral variations will not be chemically peculiar. [6]

3. Observations and processing of echelle-spectra

From the point of view of observations using modern high-resolution echelle spectrometers, stars of the A-F classes with narrow absorption lines (with a slow speed of the projection of axial rotation to the line of sight) are the most optimal objects.[7]
In 2009, a 1.5-m telescope of the RTT-150 Kazan Federal University, established in Turkey, launched a program of observations of a sample of A-F stars in a joint research program with Prof. S. Adelman (USA) and his Turkish counterparts. At the first stage, the stars included in the list of observations proposed for the observations of Prof. S. Adelman on the basis of the experience of his previous studies of A-F stars. It was been obtained high-resolution echelle spectra (R = 50000, Delta Lambda = 0.15 A) for a set of 20 stars using Andor's large-format CCD (2048 x 2048 pixels) mounted on a Coute-echelle RTT-150 spectrometer. Performing a preliminary study of the chemical composition of the 12 stars of this program (with narrow absorption lines) by means of processing and analysis of spectra by the method of atmospheric models available at the Department of Astronomy and Space Geodesy of KFU. The list of these stars is given in Table 3.1. [8]

Table 1. The main parameters of the stars from the SIMBAD database and the number of spectral lines whose equivalent widths were measured in this paper in the spectrum of the corresponding star.

HD number	Star name	V	Sp	N lines
165908	99 Her	5.07	F7 V	570
112735	-	7.28	G0 IV-V	470
130044	-	6.74	F0 IV-V	308
157329	-	6.91	F8 IV-V	213
32537	9 Aur	4.98	F0 IV-V	85
26322	44 Tau	5.40	F2 IV-V	694
147135	-	6.32	F0 III-IV	140
214994	Omi Peg	4.80	A1 IV	191
95608	60 Leo	4.41	A1m IV-V	95
187340	-	5.91	A2 III	131
72037	2 UMa	5.46	A2m IV-V	267
195295	41 Cyg	4.02	F5 I-II	408

In general, the equivalent widths of 3500 spectral lines in the echelle spectra of these 12 stars were measured. The processing of the echelle spectra was performed using a modified version of the DECH20 program by GA Galazutdinova.

4. Determination of the fundamental parameters of stars.

To calculate the contents of chemical elements, it is necessary to have an atmosphere model with predetermined parameters-the effective temperature Teff, gravity acceleration at the surface logg, the total metal content (relative to the solar one).

4.1. Determination of the effective temperature Teff.

In this paper, the effective temperature was determined by the color indices B2 - G, B2 - V1, b - y, B - V, U - B, which were calibrated in the literature in the infrared flow scale. The corresponding equations for the relationship between temperature and color indices were taken from the following articles:

The equations of the relationship between the temperature and the color indices of the Geneva medium-band system:

1)

$$\theta_{eff} = 1.626\ (B2-G) + 1.665, \qquad -0.825 \le B2-G < -0.725$$
$$\theta_{eff} = 0.81\ (B2-G) + 1.07, \qquad -0.725 \le B2-G < -0.640$$
$$\theta_{eff} = 0.458\ (B2-G) + 0.835, \qquad B2-G \ge -0.640$$

2)

$$\theta_{eff} = 2.306\ (B2-V1) + 0.917, \qquad FOR\ -0.300 \le B2-V1 < -0.160$$
$$\theta_{eff} = 0.640\ (B2-V1) + 0.632, \qquad FOR\ -0.160 \le B2-V1 < +0.730$$

3) θ_{eff}= 0.629 + 0.644*(B2-V1) + 0.065*(B2-V1)**2
4) The form of equation T = A + B*p + C*(p^2), where p is the color index
5) θ_{eff} = (U-B) - 0.72*(B-V) - 0.05((B-V)^2)

6) θ_{eff} = 0.838 + 0.501*(B2-G) - 0.030*((B2-G)^2)

In the SYMBAD database I found the values of the above color indicators and determined the effective temperature values.
Below are examples of determining the effective temperature for some stars by different color indices and by calibrations of different authors. The most deviant values were discarded. The resulting accepted (average) values of Teff obtained for the investigated stars are given in Table 1. On average, the error in determining the effective value is 100 K.

99Her Equation	color index	Color index value	T effective K
1. θ= 0.458*(B2-G)+0.835	B2-G	-0.013	6080
2. θ= 0.640*(B2-V1)+0.632	B2-V1	0.304	6094
4. T= A+B*p+C*(p^2)	B2-V1	0.304	6096
a) A= 7995 B= -7246 C= 3284			
b) A= 9865 B= -13319 C= 7483	b-y	0.356	6072
c) A= 8515 B= -5270 C= 1440	B-V	0.475	6337
d) A= 7749 B= -7911 C= 7856	B2-V1	0.304	6070
e) A= 9034 B= -12562 C= 11813	b-y	0.356	6060
f) A= 9134 B= -8600 C= 5398	B-V	0.475	6267
3. θ= 0.629+ 0.644*(B2-V1)+0.065*((B2-V1)^2)	B2-V1	0.304	6065
5. θ= (U-B)-0.72*(B-V)-0.05*((B-V)^2)	(U-B), (B-V)	1.207, 0.475	6154
6. θ= 0.838+0.501*(B2-G)-0.030*((B2-G)^2)	B2-G	-0.013	6065

4.2 Determining the value of log g

The determination of the acceleration due to gravity in this work was carried out taking into account the fact that for all stars there are high-precision values of trigonometric parallaxes measured by the satellite GIPPARCOS.
Knowing the effective temperature and luminosity of the star, we can determine log g by the formula:
log g = 4 * log (Teff) + 0.4 * Mv + log (M / Mo) -12.51,
where the constant value 12.51 reflects the solar values Teff = 5770K,
log g = 4.44, Mv − V + 5 + 5 * logπ
The trigonometric parallax values were taken from the SYMBAD database

Table 2: The trigonometric parallax values were taken from the SYMBAD database

Star	V	π, arcsec	Mv	log g
99 Her	5.07	0.064 +/-0.0005	4.09	4.4

5. Conclusion:

Within the framework of this work, echelle spectra of star 99Her of spectral classes A-F has been proceed, obtained with a 1.5-meter telescope PTT-150. As a result of processing the spectra, the equivalent widths of unbend and weakly blended absorption lines were measured, 350 lines this star. The measured equivalent line widths can be used for other studies of the atmospheres of the A-F stars. For 99Her star, the fundamental parameters of atmospheres-the effective temperature, the acceleration of gravity, the microturbulent velocity, were determined, and the content of the chemical elements was calculated (on average for 10-20 chemical elements for each star.

Comparison with the published calculations of other authors has shown the reliability of the results obtained. Within the accuracy of the determination of the chemical composition, our results agree with those published.

Stars with temperatures of 8000-9000 K show a significant excess of elements heavier than iron and some deficiency of light elements. These results, apparently, are a consequence of the effects of light pressure and gravitational diffusion of atoms and ions in atmospheres of calm and radiative atmospheres of A-stars. Unfortunately, the chemical composition of the atmospheres of our star does not reflect the real chemical composition of the substance from which originated.

References

[1] O.V. Abramova, "The Chemical Evolution of Gas in Galaxies." The role of accretion and outflow of gas "edited by D.Sc. Zasova A.V. 2018

[2] L.I. Mashonkina, V.V. Shimansky, N.A. Sakhibullin "Non-LTE effects in the spectral lines of NaI in the atmospheres of stars of different types", Astronomical Journal, 2000, vol. 77, No. 12, p. 1-16

[3] S.J. Adelman, I.H. Cay, H.G. Tektunali, A.F. Gulliver, A. Teker «Elemental abundance analyses with DAO spectrograms: XXXI. The early F supergiants ν Her (F2 II) and 41 Cyg (F2 Ib-II) ". Astron. Nachr. , 2008, AN 329, No. 1, p. 4- 9.

[4] S.J. Adelman, H. Caliskan, D. Kocer, I.H. Cay, H.G. Tektunali «Elemental abundance analyses with DAO spectrograms - XXIII. The superficially normal stars 28 And (A7 III) and 99 Her (F7 V), Mon. Not. R. Astron. Soc., 2000, No. 316, p. 514-518

[5] S.J. Adelman, O.I. Pintado, F. Nieva, K.E. Rayle, S.E. Sanders, Jr. "On the effective temperatures and the surface gravities of the superficially normal main sequence band B and A stars", Astronomy And Astrophysics, 2002, v.392, p.1031-1037

[6] F. Castelli, R. Kurucz, "New Grids of ATLAS9 Model Atmospheres", IAU Symp. No 210, Modeling of Stellar Atmospheres, eds. N. Piskunov et al. 2003, poster A20

[7] A.F. Gulliver, S.J. Adelman, T.P. Friesen "A spectroscopic atlas of Pegasi (A1 IV) λλ 3826- 4882", Astronomy And Astrophysics, 2004, v.413, p. 285-291

[8] I.H. Cay "A spectroscopic analysis of the metallic-line star 2 UrsaeMajoris (A2m)", The A-Star Puzzle, Proceedings IAU Simposium, 2004, No. 224, p. 673-682.

[9] C. Soubiran, P.Girard "Abundance trends in kinematical groups of the Milky Way's disk", Astronomy And Astrophysics, 2005, v.438, p.139-151

[10] W. Zima, H. Lehman, Ch. Sturtz, I.V. Ilyin, M. Breger «High-resolution spectroscopy of the δ Scuti star 44 Tau: photospheric element abundances and mode indenfication», Astronomy And Astrophysics, 2007, v.471, p.237-245

IOP Conf. Series: Materials Science and Engineering 1095 (2021) 012005 doi:10.1088/1757-899X/1095/1/012005

Effect of Reinforcement Materials and Immersion Mediums on Mechanical Properties of Epoxy Composites

Mohanad R. Al-Salmani[1], Ahmad H.M.Alfalahi[2]

[1]Physics department ,College of Education for pure Sciences, University of Anbar, Ramadi, Iraq
[2]Physics department ,College of Education for pure Sciences, University of Anbar, Baghdad, Iraq

[1]**Corresponding author:** moh200089@gmail.com ; moh200098@uoanbar.edu.iq

Abstract : In this research, a non-reinforced epoxy and its composites which reinforced by fiber glass, polypropylene fiber and micro-silica oxide powder, with volume fraction (20%) were prepared , using a hand lay-up molding method . Impact strength, Surface hardness, and Bending strength tests were carried out for the specimens before and after immersion in tap water and ethanol. Before immersion, the results showed that adding fiber glass and polypropylene fiber to the polymer improved the mechanical properties while silica particles decreased it . On the other hand water absorption of the substance led to a negative effect on the mechanical properties of the composite material. The effect of ethanol was the highest.

Keywords: Epoxy, Fiber glass, Polypropylene, Silica Oxide, immersion mediums, mechanical properties.

1. Introduction.

Development of modern science and technology require the need to find new materials instead of metals and ceramics, these materials are called composites which have a light weight , high strength, electrical and thermal insulation, in addition to their resistance to the environmental conditions, to be suitable for industrial uses [1]. Composite materials consist of matrix material, which represents the continuous phase, reinforcement material {Reinforcement Phase} which stands most of the stress applied on the composite material, leading to improve its mechanical properties, and the third phase is the interface region located between the matrix and reinforcement material [2]. The composite material resistance for the applied stress mainly depends on the elasticity modulus and the strength of the reinforcement fibers and particles. The amount of reinforcement materials in the composites is known by the volume fraction [3]. The bending resistance of the material is its ability to endure the vertically applied bending forces on its longitudinal axis. The stress resulting from applying the bending load combines two types of stress, compression and tensile stress [4].

2. Experimental Procedures .

2-1. Samples preparation : Epoxy resin type (sikadur 52) made by Sika company , was used as matrix material .It is a thermoset polymer, transparent viscous liquid, adhesion ,little shrinkage, and it turns to hard material by adding a hardener with a mixing ratio 1:2 in laboratory temperature. Fiber glass, polypropylene fibers, and micro-silica oxide were used as reinforcement materials with volume fraction 20% . The weight of reinforcement materials was calculated according to the following relations.

$$Wc = Wf + Wm \text{ ....(1)}$$

$$Vf = \frac{1}{1 + \left(\frac{(1-\psi)}{\psi}\right) * \left(\frac{\rho f}{\rho m}\right)} \text{ ....(2)}$$

$$\psi = (wf / wc) * 100\ \% \text{ ....(3)}$$

ψ : Weight fraction

wc ,wf, wm: weight of the composite, reinforcement, and matrix materials respectively(g).

ρf, ρm: Density of reinforcement and matrix material respectively(g/cm3).

Vf: Volumetric fraction (%)

Hand Lay-Up molding method was used in preparing the following samples :-

1-Unsupported epoxy , 2- EP+ F.G , 3- EP+ P.P.F , 4-EP+ SiO_2

-Impact test: The following tests were done for the samples**:**

Charpy test, type shock monitor, was used to determine the impact strength of the material.

Specimens were cut according to the standard specifications (ISO-179), impact strength was calculated using eq.4.

$$I.S = \frac{E_c}{A} \text{ ... ... (4)}$$

I.S : impact strength (MPa) , E_c : energy of fracture (J) , A : cross section area (mm^2)

-Hardness test : To investigate the surface hardness of the composite materials, the duroment hardness (type Shore – D, TH210 Equipment device), was used to perform a hardness test .

-Bending Test : Three points bending test was done to investigate the ability of the composite material to resist the applied load, which illustrate the nature of the material .

Flexural strength of the samples is calculated according to eq.5.

$$F.S = \frac{3PL}{2bd^2} \text{...........(5)}$$

F.S: flexural strength (N/mm^2).

P: maximum load (N) , L: the space between the two pillars (mm)

b: sample width (mm). , d: Sample thickness (mm)

3. Results and Discussions

Impact strength : Tables (1 & 2) and figures (1 & 2) provide that reinforcing the polymer by fibers increase the resistance of composite material to impact , and that was in case of fiber glass higher than polypropylene fibers, while reinforcing by silica oxide reduced the impact resistance. Fibers bear the major part of the stress applied on the composite material more than particles, and prevents cracks diffusions and act as bumpers of fracture and increase the material cohesion and thus increases the impact strength [5]. On the other hand, adding silica particles to the polymer gave it the brittleness properties because of ceramics particles were known so as a brittle material and it may provide points to focus stresses and micro cracks positions and their diffusions in composite material which case impact reduction [6].

After immersion in water and ethanol, a decrease in the impact resistance of all specimens was observed. Diffusion of water and its penetration through the composite material, especially through the polymer and voids created during the manufacturing process causes swelling and formation of stresses within the material, which leads to weak it and failure. On the other hand, absorption of water and ethanol causes plasticization of the matrix material as a result of the breakdown of the hydrogen bonds between the polymeric chains, the material loses its ductility and becomes brittle and thus its resistance to shock decreases [7]. Ethanol effects was higher than the water.

Surface Hardness : Tables (3 & 4) and figures (3 & 4) show that adding fibers and silica oxide to the polymer improved the surface hardness of the composite material specially the fiber glass and silica particles because of their higher hardness and ability to endure stresses and plastic deformation [8].

After immersion in water and ethanol, surface hardness of the polymer and composites was decreased with the length of the immersion period. Water and ethanol penetration into the interface region and the voids leads to a breakdown of the bonds between the polymer chains thus weakening the cohesion of the composite material and its resistance to scratching and the surface hardness decreases.The ethanol effects was greater than that of water because ethanol is a corrosive chemical solutions[9],[10].

Table.3: variation of surface hardness vs immersion time in H_2O

Flexural Strength: Tables (5 & 6) and figures (5 & 6) illustrate that adding the fibers to the polymer increased the resistance of the composite material to bending, specially the glass fibers, which are characterized by high strength. On the other hand, adding silica particles decreased the flexural strength. The silica particles are characterized by brittleness, which makes the composite material brittle thus reducing the flexural strength [11].

After water and ethanol immersion, the flexural strength decreased in varying degrees for all samples in proportion to immersion time , as a result of the penetration of solutions inside the composite material and their negative effect. The effect of ethanol was greater than that of water, because ethanol is a corrosive chemical solutions [9],[10].

4. Conclusions.

From the obtained results , the following can be concluded**;** Supporting the polymers with glass and polypropylene fibers improve the impact strength, surface hardness and flexural strength of the composite material**;** While micro-silica oxide particles decrease the impact and flexural strength of the composite material**;** Exposure of the polymer and composite to water and ethanol negatively affects the mechanical properties of the material depending on the type of solution , the duration of immersion , and the type of reinforcement material .

References

[1] Autra K. Kaw, 2006," Mechanics of composite materials " , 2ed edn.,Taylor & Francis group. LLC.

[2] Clyne T.W. and Hull D., 2019," An introduction to composite materials " 3ed edn., Cambrige university press.

[3] William D. Callister ,Jr.,2007, "Fundamental of Materials Science and Engineering ", 7th edn., John Wiley & Sons ,Inc.

[4] Hassan T. and Balkees M. Al-Dabbagh,(2013). "Study the Effect of Coupling Agents on the Physical Properties of Polymer Composite Material",MSc thesis, University of Technology. Baghdad, Iraq.

[5] Wiphawee Nuthonga , Putinun Uawongsuwanb *, Weraporn Pivsa-Art a , Hiroyuki Hamada., 2013," Impact Property of Flexible Epoxy Treated Natural Fiber Reinforced PLA C y p composites", Energy Procedia 34 ,839 – 847

[6] Ahmad H.M. ,and Ja'afar K. Ch.,2018," Study of Wear Rate and Some Mechanical Properties for Nanocomposites [Blend /SiC-PZT] ", Journal of Engineering and Applied Sciences ,vol.13.no.23 .pp.9844-9848

[7] Balkees M.D. and Ahmad H.M.,2012," Effect of liquids absorption on some mechanical and thermal properties for polymeric composites " proceeding of the international conference on engineering and information technology , Toronto Canada , ISBN: 978-1-77136-064-7, pp.1-6

[8] - Gagani, A. I., Monsås, A. B., Krauklis, A. E., & Echtermeyer, A. T. (2019). The effect of temperature and water immersion on the interlaminar shear fatigue of glass fiber epoxy composites using the I-beam method. Composites Science and Technology, 181, 107703. doi:10.1016/j.compscitech.2019.107703

[9] Ahmad H.M. Al-falahi , Balkees M.D. Aldabagh ,& Bara'a J. Rashad,2018, "Effect of HCL acid on the physical properties of Hybrid /Epoxy nanocomposites ", proceeding of first international and third scientific conference , college of science /university of Tikrit , Iraq ,Dec.2018 /p4, pp.173-176

[10] Ahmed W.I., 2013, "Study The Effect Of Short Carbon Fibers On Properties Of Epoxy Composite Material",Ms.C. thesis, , College of Education for Pure Science , University of Anbar, Iraq

[11] Aseel M.A.,2010," Study of mechanical properties for polyester reinforced by date palm with maleic, NaOH, and HCL Chemical Treatments " Journal of Engineering and Technology , Vol.28,no.11

Table 1: variation of impact strength vs immersion time in H_2O

Sample	Impact Strength (KJ/m²) Immersion Time (Week)				
	0	2	4	6	8
EP	24.443	21.222	18.052	14.882	11.7123
EP+F.G	46.033	36.181	31.549	22.704	20.98
EP+P.P.F	24.972	15.356	12.348	12.275	8.066
EP+SiO_2	8.384	6.687	5.184	4.825	4.721

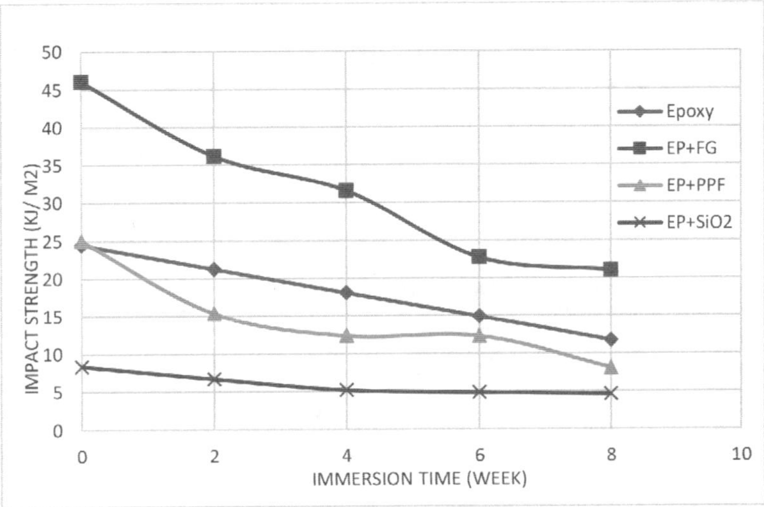

Fig.1: variation of impact strength vs immersion time in H_2O

Table 2: variation of impact strength vs immersion time in ethanol

Sample	Impact Strength (KJ/m²) Immersion Time (Week)				
	0	2	4	6	8
EP	24.442	19.822	15.252	10.682	6.1132
EP+F.G	46.033	32.673	23.550	19.70	15.027
EP+P.P.F	24.972	17.877	15.514	14.516	12.903
EP+SiO_2	8.384	7.503	5.706	4.825	5.5

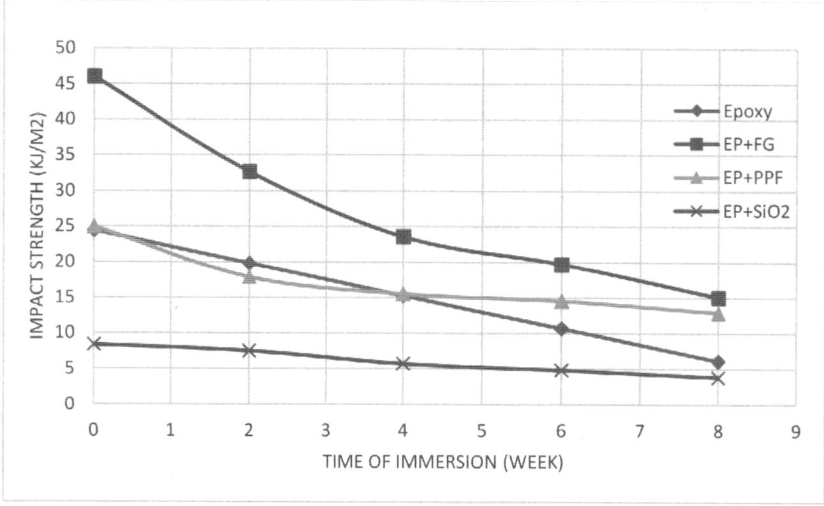

Fig.2: variation of impact strength VS immersion time in ethanol

Table.3: variation of surface hardness vs immersion time in H_2O

Sample	Hardness				
	Immersion Time (Week)				
	0	2	4	6	8
EP	79	74	70.3	66.60	62.8
EP+F.G	83	81	81	80	78.1
EP+P.P.F	78	77.6	77	77	76.9
EP+SiO_2	83.7	82.7	82.2	82.2	82

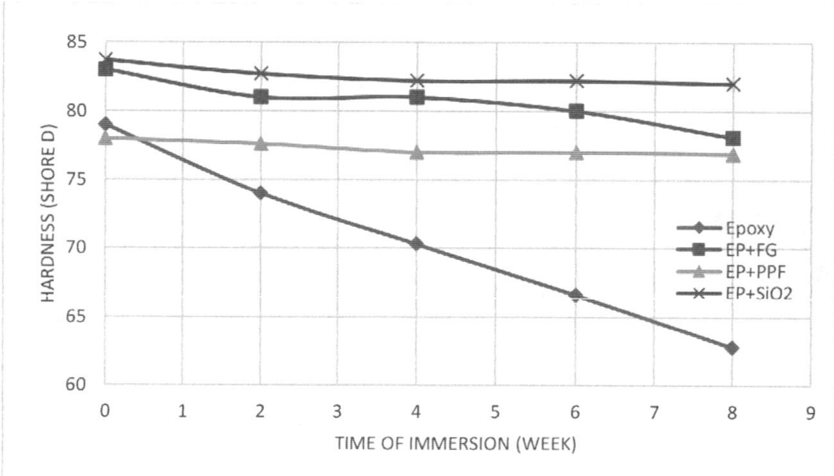

Fig.3: variation of hardness strength vs immersion time in H_2O

Tabl.4: variation of surface hardness vs immersion time in Ethanol

Sample	Hardness				
	Immersion Time (Week)				
	0	2	4	6	8
EP	79	69.4	60.5	51.68	42.8
EP+F.G	83	74.2	73.5	68.8	68
EP+P.P.F	77	60.5	59.1	64	62.7
EP+SiO_2	83.7	69	66	65.8	65.2

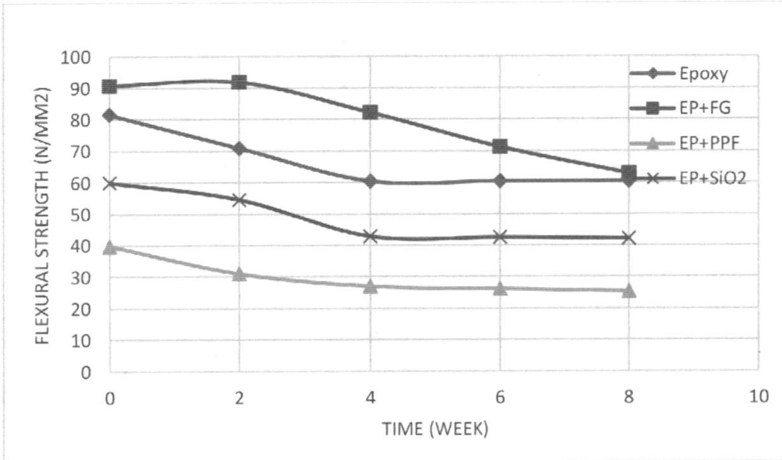

Fig.4: variation of hardness strength VS immersion time in Ethanol

Table 5: Variation of flexural strength vs immersion time in H_2O

Sample	Flexural Strength(N/mm2)				
	Immersion Time (Week)				
	0	2	4	6	8
EP	81.5	70.9	60.414	49.05	36.68
EP+F.G	116.185	100.338	91.894	81.7	77.664
EP+P.P.F	88.63	46.941	45.150	39.86	39.278
EP+SiO_2	49.165	40.746	38.392	35.375	33.660

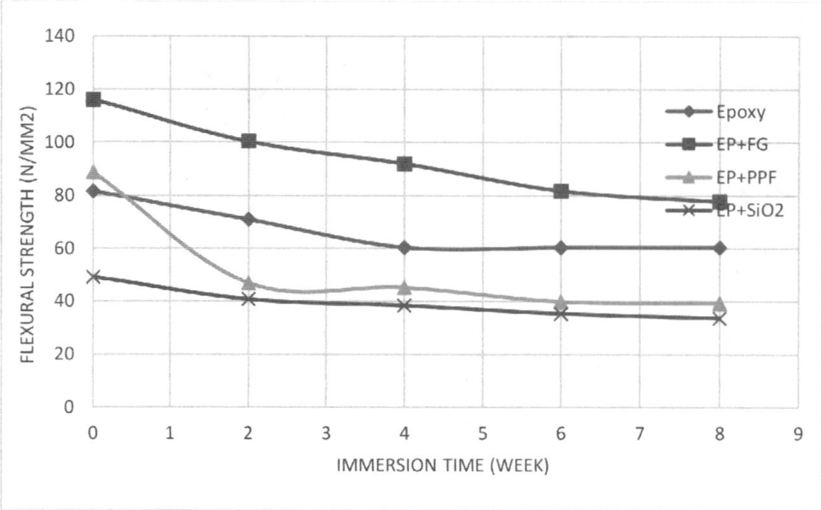

Fig.5: Variation of flexural strength vs immersion time in H_2O

Table.6: Variation of flexural strength vs immersion time in ethanol

Sample	Flexural Strength(N/mm2)				
	Immersion Time (Week)				
	0	2	4	6	8
EP	81.5	11.53	11.53	11.53	11.53
EP+F.G	116.185	90.66	59.059	45.528	31.967
EP+P.P.F	88.63	49.165	35.375	33.584	11.892
EP+SiO_2	49.165	30.732	20.7	17.542	9.281

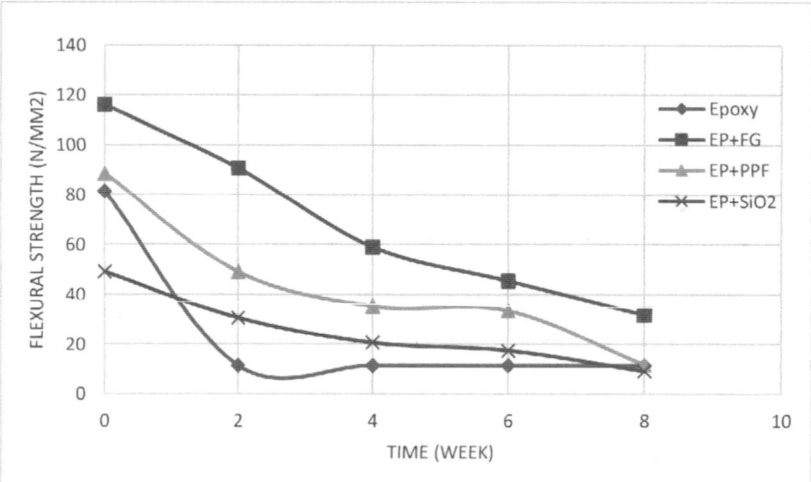

Fig.6: variation of flexural strength VS immersion time in ethanol

Influence Different Substrates on the Sensing Performance of Copper Oxide Thin Films Synthesis via the Hydrothermal Method

Ahmad Z.Al-Jenaby [1,3], Asmiet Ramizy[2] and Abdul-Majeed E Al-Samarai[3]

[1]Ministry of Education, Direction of Education in AL-Anbar
[2]Department of Physics , College of Science, University of Anbar, Anbar, Iraq
[1,3]Department of Physics, College of Education for Pure Sciences, University of Tikrit, Tikrit, Iraq

*E-mail : azidan652@yahoo.com

Abstract We report the synthesis of CuO nanostructures on silicon(Si) and porous silicon (PS) substrate using hydrothermal method as seed layer and different reaction time (12 and 24) hour respectively. structure and sensing properties analysis of these nanostructures were studied using X-ray diffraction (XRD), (AFM) and FE-SEM analysis . all the obtained peaks in the XRD pattern are well matched with the monoclinic phase of CuO thin film prepared on silicon and pours silicon substrate with crystal sizes (25.2 to 11.6) nm and (21 to 9.9) nm for CuO/Si and CuO/PS respectively. Topography analysis studied by (AFM) and reveals that the grain size, roughness surface and root mean square of the prepared thin film decreases with the increase in the hydrothermal reaction time in both cases CuO / Si and CuO / PS. FESEM analysis, showed the morphological of film prepared at seed layer as seed-like structure, and change to nanosheets structure with increasing reaction time. EDX analysis revealed the chemical composition and elemental state of CuO nanoparticles. Gas sensor performance showed that thin films have good sensitivity properties but thin films deposited on porous substrates are better and amount to about 186%.

Keywords: Structural and Sensitivity, Hydrothermal.

1. Introduction

p-type cupric oxide (CuO) semiconductor has an energy gap 2.8 ev at bulk, and monoclinic crystal structure with lattice parameters of cupric oxide [1] are a = 4.6837A°, b = 3.4226A°, c = 5.1288A°, $\alpha = \gamma = 90°$, $\beta = 99.54°$. It is an interesting material due to the applications promising for electronic and optical devices, such as electrochemical cells [2], gas sensors [3] magnetic storage media [4], solar cell [5], field emitters [6], nanoscale devices for catalysis solar cells, etc. [7,8] Among these CuO nanostructures, the formation of CuO nanowires and nanoparticles has been studied using several techniques, such as spin coating [9], hydrothermal[10], self-catalytic growth [11], and solvothermal [12] recently, CuO nanoparticles were formed by using a sol–gel technique [13], one-step solid state reaction [14], electrochemical method [15], and thermal decomposition of precursors [16] In the present study, we have synthesized CuO nanoparticles using different structures Si and PS/Si by a hydrothermal process. The aim of this research is to study the effect of different substrates on the structures and sensitivity of CuO synthesis by hydrothermal method. Due to increased environmental awareness, the need for highly efficient combustion engines, and controlled and controlled industrial processes that monitor the level of hazardous gases in the environment, has led to the search for an effective gas sensor[17]. A good gas sensor should have high sensitivity and short response time, stability, and good selectivity at low-level operating temperatures[18]. The sensitivity and selectivity of gas sensors can be improved by changing some parameters. The aim of this study, is to the synthesize CuO nanostructures by hydrothermal method with different reaction times, in order to get CuO Nanosheets and study the structural and sensing properties of CuO nanostructures on Si and PS substrates with changing hydrothermal reaction time.

2. Experimental details

All materials used in the experiment have a high degree of purity. CuO nanostructures were prepared by hydrothermal method with Sodium hydroxide (NaOH) as precipitating agent as well as hexamethylenetetramine ($C_6H_{12}N_4$) as a catalyst to increase the number of OH ions. PS layers were formed through the Photo-electrochemical etching (ECE) of n-type <111> oriented silicon substrate. In a typical CuO synthesis, 1.3 g of $CuCl_2$;2 H_2O was dissolved in 40 mL deionized water under constant stirring for 60 minutes until a homogeneous blue colour solution is obtained . Then 10 ml of an aqueous solution of sodium hydroxide solution is added. The aqueous solution of NaOH was added to the above solution gradually and under constant stirring. The pH value of the solution was maintained to be 3.75 - 4 . The resultant solution was transferred to the stainless steel autoclave and hydrothermal synthesis was carried out for about 120°C for 3h. The obtained black precipitate was washed several times with deionized water and acetone to remove impurities. The final product is dried at 80°C for 2h inside the oven . Then we use spin coating technique 2500 revolutions / minute to form the seed layer on all Si and PS substrates and dry the thin films for 1 h inside the oven, then the films are annealed at 375 ° C for 3 h. The seed layer is placed inside a Teflon autoclave and heated at 120 ° C with a change in hydrothermal time (12 and 24) hours. Then, the substrates removed from the autoclave, cleaned by distilled water and all the samples were annealed for 3h at 375 °C. Structure properties, including crystal size and crystal structure, were studied by X-rays Diffraction System (BRUKER AXS), whose radiation source was Cu (kα) and wavelength 1.5406 A°. The Film surface morphology tested via AFM, SPM model AA3000, and FESEM type (FEI Verios 460L) From the Supra 50VP device. A UNI-T UT81B laptop computer an digital multimeter were used to record the change in sensor resistance when exposed to air-NH_3 gas mixing ratio.

3. Results and discussion

3.1. Structure study

3.1.1. X-ray diffraction

X-ray diffraction (XRD) studies are performed to study the crystalline properties of CuO thin films. Figure a-1 illustrates the XRD patterns of CuO thin films deposited on Si (111) substrate with different reaction times. The XRD patterns of the CuO thin films deposited on the Si(111) substrate indicate the formation of the phase of CuO at the preferred orientation (002), and that all the thin films are polycrystalline with the appearance of the peak for $Cu(OH)_2$ indicating the effect of the addition of $C_6H_{12}N_4$ which leads to an increase of OH ions for all samples. X-ray diffraction peaks corresponding to 2θ =35.5°, 38.7°, 48.88° and 61.68° are of (002), (1 1 1), (-202) and (-113) for all these samples could be perfect indexed to CuO monoclinic (JCPDS Card No. 05-0661). Moreover, the expansion of the diffraction peaks is an indication that the prepared material is in the nanometer range. Figure b-1 , illustrated XRD patterns of samples CuO/PS with a change in the hydrothermal reaction time. The peaks at 2θ= 32.58 °, 35.55 °, 38.79 °, 48.76 °, 58.3 °, 61.59 °, 66.2 °, and 68.1° are of (110), (002), (111), (-202), (202), (-113), (-311), and (200) planes respectively, with the monoclinic structure of CuO (JCPDS Card No. 41-0254) This result is in agreement with that of [19,20] The peaks at 2θ= 28.4° attributed porous silicon while 75.1° attributed to the impurity Cu(OH)2 indicating the effect of the addition of $C_6H_{12}N_4$ [21] The time reaction increased did not change the direction of crystalline growth for the dominant planes, the intensity of XRD peaks decreases accompanied by an increase in the FWHM (full width at half maximum). Finally this confirms that the copper oxide has been cultivated on Si and PS substrates successfully, thus the monoclinic CuO phase is the same in the case of both substrates grown on Si and PS substrates. The average crystallite size of the deposited thin films is calculated from the line broadening using Scherrer formula [22]:

$$D = K\lambda/\beta\cos\theta \quad \ldots\ldots\ldots\ldots \quad (1)$$

Where λ represents the wavelength of the x-ray radiation 0.154nm, β_{hkl} Full width half of the maximum diffraction peak (radians), θ is the Bragg angle and (k) is form factor equal to 0.9. The result shows that the average crystal size of CuO decreases when the hydrothermal reaction time increases about (25.2 to 11.6) nm and (21 to 9.9) nm for CuO/Si and CuO/PS respectively, which could be the result of crystals formation and stabilization[23]. Thus the result indicates that nanocrystalline are the domains.

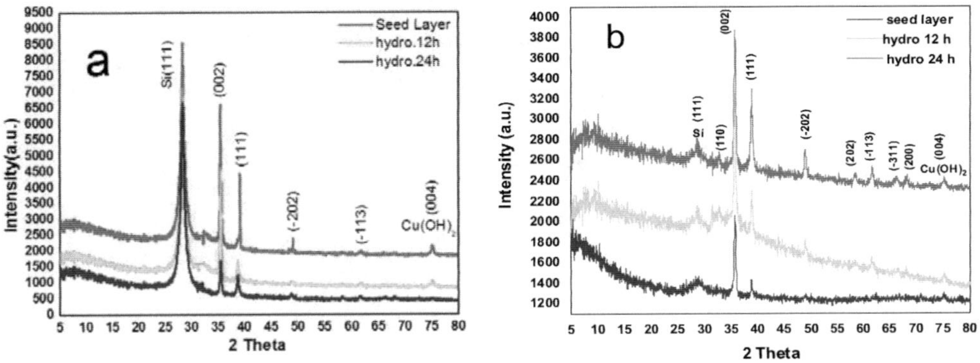

Figure 1. x-ray diffraction patterns of (a)CuO/Si (b) CuO/PS films at seed layer, 12h, and 24 h.

3.1.2 AFM Surface Topography

Figure 2 and 3 showed the images of the AFM topography of the CuO film deposited on a Si and PS with seed layer and a change reaction time. it showed the total surface topography and the appearance of the grains formed on the surface of the film, As XRD results differ from AFM, XRD determines the crystalline size and depth inside the structure, while AFM determines grain size on the surface [24]. The effect of reaction time on film surface roughness and root mean square (RMS) ware also studied. Surface roughness and (RMS) decreases with reaction time. Grain size of films decreases with increasing reaction time and it was found to be in good agreement with the size of the crystals obtained from XRD Data for all samples, similar reports describe such behaviors for other materials film [25]. Table 1, it showed that the Grain size, roughness and RMS of CuO/Si and CuO/PS respectively .

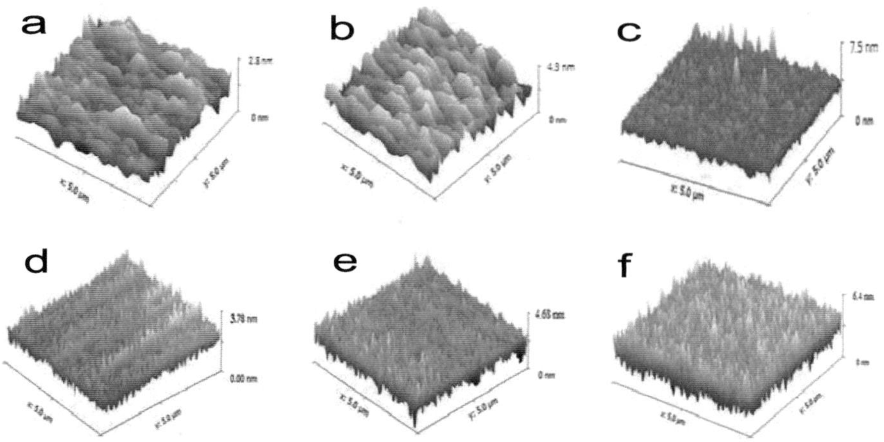

Figure 2 . 3-D AFM micrographs of(a-c) CuO/Si, (d-f) CuO/PS films at seed layer, 12h and 24 h.

Table . 1 AFM parameters of CuO/Si and CuO/PS films at seed La., (12 and 24) h .

Sample	Ave. diameter (nm)	Roughness(nm)	R.M.S (nm)
Seed La./ Si	61.7	3.91	3.68
12h/Si	49.61	3.32	3.00
24h/Si	35.60	2.46	2.40
Seed La./ PS	48.01	3.32	4.27
12h/PS	36.97	2.81	3.22
24h/PS	19.83	2.01	2.6

3.1.3 Morphological Properties by FESEM

Figure 3 and 4 shows a FESEM image of the CuO morphology of CuO at seed layer and (12 and 24)h. Morphological features have power Effect on the properties of the sensor for the nanostructure. That it can be seen that hexamethylenetetramine has a significant effect on CuO growth using NaOH as a compound a showed the formation of a CuO nanosheets nanostructure [26]. The hydrothermal synthesis reaction time often affects the shape, volume distribution, as well as clumping prepared material. Figure 3a and 4a show the seed-like CuO nanostructures it was obtained without the hydrothermal process, whereas with hydrothermal the CuO nanostructures are sheet-like with a change in reaction time, this could be due to the influence of pressure and temperature , as shown in Figures 3b and 4b. The chemical composition of the nanosheets was analyzed by energy dispersive X-ray spectroscopy (EDX), where the spectrum appears in Figures d-3 and d-4 which reveal the formation of the nanosheets of Cu and O only, and the atomic ratio of Cu to O is very close to the ratio of 1: 1, which corresponds to a stoichiometric ratio of CuO. Figure 5 shows the cross section of the porous layer which was about 26 μm, while the CuO film

thickness was 65.13 ± 5 nm. It was observed that the film consisted of nanoparticle sizes with a slightly uniform coverage of the substrate surface with randomly oriented morphology and particle size 35 nm range for different reaction time.

Figure 3. FESEM images of CuO/Si nanosheets: (a) seed layer, (b) with 12h, (c) with 24h, and (d) energy-dispersive X-ray (EDX) spectra of CuO.

Figure 4. FESEM images of CuO/PS nanosheets: (a) without hydrothermal, (b) with 12 h, (c) with 24 h, and (d) energy-dispersive X-ray (EDX) spectra of CuO.

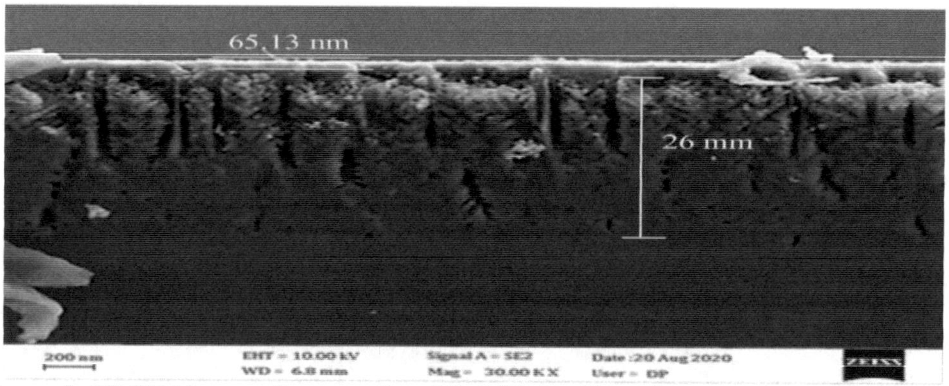

Figure 5. FE-SEM images of cross section of CuO/PS thin film and cross section of PS.

3.2 Sensing performance

3.2.1 gas sensing of CuO Nano film on Si

Figure 6 shows the resistance of the CuO seed layer based on Si substrate as a function of time with on/off gas valve exposed to 47.7 ppm of NH3gas at RT, 75 and 150 °C. Figure 7 and 8 showed the resistance of the CuO with (12 and 24)h of reaction time, and showed that the sensitivity behavior of the CuO gas sensor increased with increasing temperature. Sensitivity improved dramatically as the heat and pressure inside the autoclave led to a decrease in the grains size, thus increasing the area of gas reaction with the oxide surface in the metal oxide structure. The reason for the resistance change is due to the interaction of the sample surface with a gas, the oxygen absorbed on the surface of metal oxide and the adsorption of ion gases molecules [27] An increase in sensitivity was also observed with an increase in the operating temperature, because the increase in the operating temperature increases the adsorption of the gas on the surface of the oxide, thus increasing the reaction process on the surface, and this helps to recompenetion and thus the number of holes decreases and thus the oxide surface resistance increases.

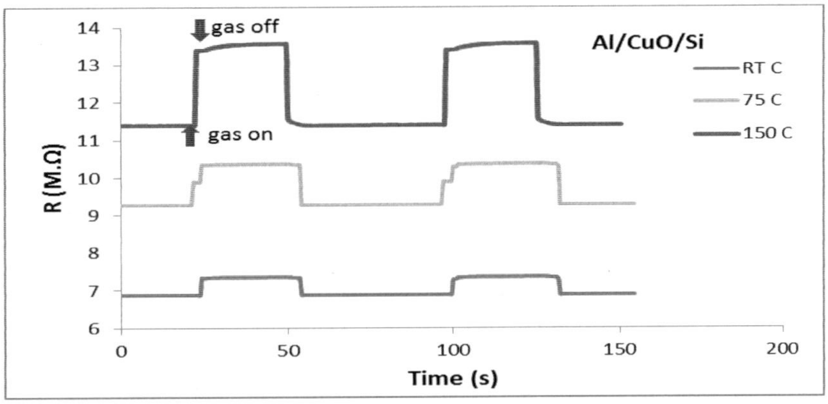

Figure 6. Resistance of Al/CuO/Si seed layer as a function of time for NH_3 gas.

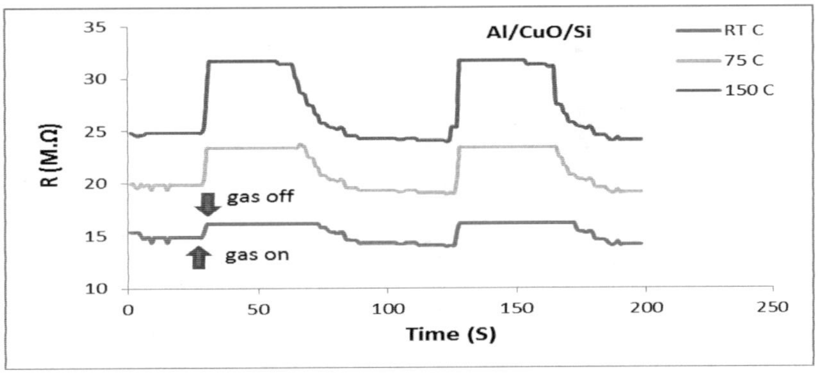

Figure 7 . Resistance of Al/CuO/Si at 12 h as a function of time for NH_3 gas

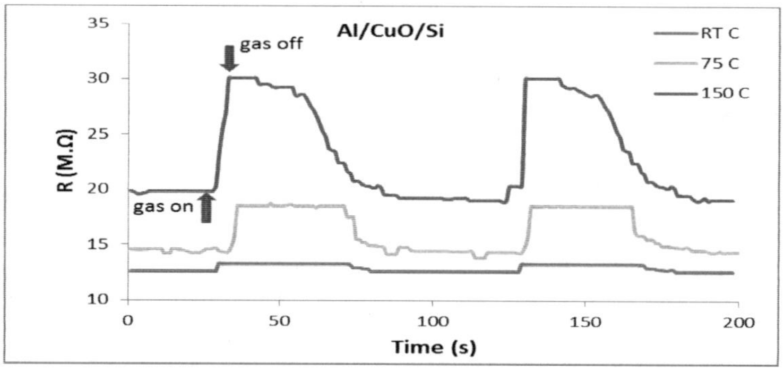

Figure 8. Resistance of Al/CuO/Si at 24 h as a function of time for NH_3 gas

3.2.2 gas sensing of CuO Nano film on PS

Figure 9 shows the resistance of the seed layer CuO/PS substrate as a function of time with on/off gas valve exposed to 47.7 ppm of NH_3 gas at RT, 75, and 150°C, The results showed that the resistance decreased when Films exposed to NH_3 (gas oof), then the resistance value increases when it is shed gas (gas on) [28]. Figure 7 showed the sensitivity increases about more than doubled 16% compared to CuO / Si seed layer about 7.2% at RT. This indicates how porous silicon affects the NH_3 sensor, and this is demonstrated by changing the resistance behavior the response and recovery time improve when the deposited on porous compared with the sample on silicon as table2.

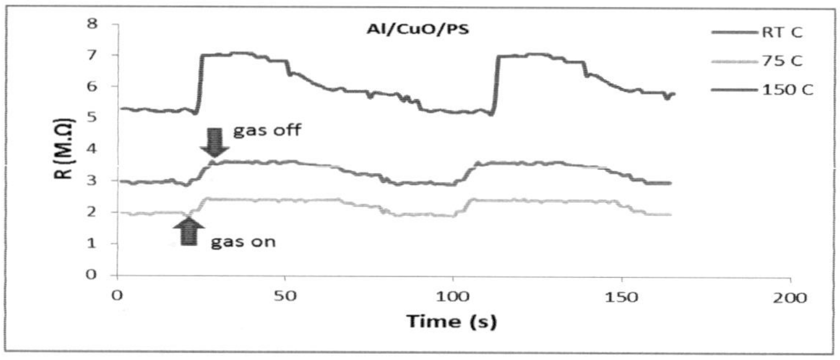

Figure 9. Resistance of Al/CuO/PS seed layer as a function of time for NH_3 gas.

Figure 10 and 11 illustrate sensitivity as a function of operation The temperature of CuO with variation in reaction time, which is deposited on PS substrates. Gas sensitivity tests were performed (RT, 75 and 100)°C . Results at 12h reaction time

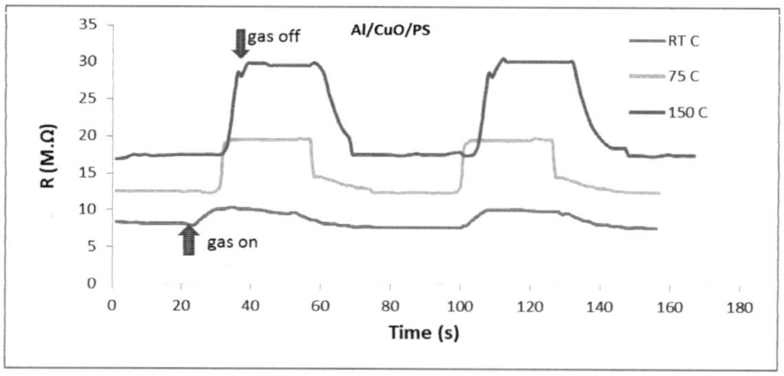

showed the sensitivity increases with increasing the temperature, this means improved film sensitivity with an increase in the operating temperature, this may be attributed to increase the surface reaction rate of the target gas due to the porosity possessed by the PS substrates. But at a reaction time of 24h a high increase in sensitivity occurred with an increase in the operating temperature, The optimum temperature was observed at the temperature is 150°C. Maximum sensitivity to NH3 was noted the film upon this reaction was found to be 186% as showed in the results Listed in Table 2, this is a very good result indicating that PS substrates good for detecting NH3 gas and is better than silicon substrates, this may be because the surface was able to oxidize the gas intensively and thermodynamic stable [29,30].

Figure 10 Resistance of Al/CuO/PS at 12 h as a function of time for NH_3 gas

Figure 10 . Resistance of Al/CuO/PS at 12 h as a function of time for NH_3 gas.

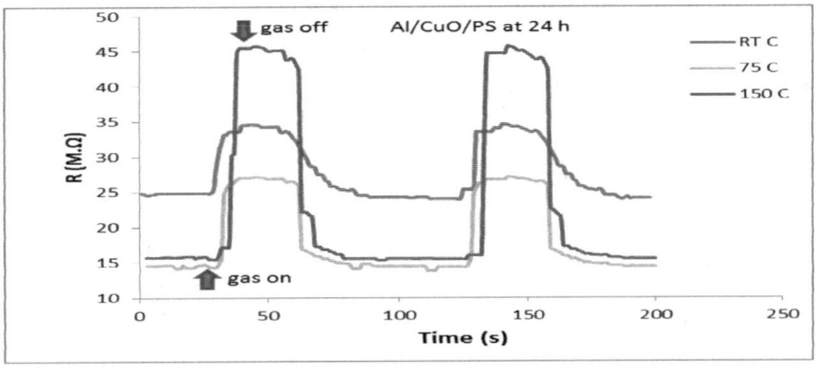

Figure 11 . Resistance of Al/CuO/PS at 24 h as a function of time for NH_3 gas.

Table 2. Response time, recovery time and sensitivity % of CuO/Si and CuO/PSwith seed layer and different reaction time (12 and 24) h.

Sample	Response time (s)			Recovery time (s)			Sensitivity %		
	RT	75	150	RT	75	150	RT	75	150
Seed La. /Si	26.5	27.6	24.7	19.8	22.5	23.4	7.2	11	18.9
12h CuO/Si	21.6	25.2	21.6	22.5	21.6	24.3	8.5	18	28.0

24h CuO/Si	24.3	20.7	19.8	19.8	19.7	20.7	10	28	41
Seed La. /PS	22.5	23.3	22.7	21.6	20.5	22.4	16	11	31
12h CuO/PS	20.1	23.1	23.2	23.5	21.1	23.3	23	57	69
24h CuO/PS	21.5	21.3	19.2	19.2	19.9	21.4	39	83	186

4.Conclusion

In this paper, an easy and inexpensive hydrothermal method that is used for synthesis is illustrated Thin films CuO/Si and CuO/PS at the seed layer and different reaction times (12 and 24) h . From the XRD analysis , it confirmed that the synthesized CuO nanosheets had monoclinic CuO structures and the crystal size decreases due to increasing the reaction time of both substrates to 11.6 and 9.9 nm, respectively. AFM topographer indicates grain size at the nanoscale, and decreases with increasing reaction time. FESEM confirms that the morphology of all the prepared thin films changes from seed-like structure to nanosheets with increasing reaction time. Energy dispersion (EDX) spectrum confirms the presence of Cu and O in the prepared materials. The results of the gas sensing performance showed that the thin films have good sensitivity properties, but thin films deposited on porous substrates CuO/PS are of better and higher performance with a sensitivity of about 186%.

References

[1] J. Chen, S.Z. Deng, N.S. Xu, W.X. Zhang, X.G. Wen and S.H. Yang 2003 Temperature dependence of field emission from cupric oxide nanobelt films Appl. Phys. Lett. **83** 746–748.

[2] A. Chowdhuri, V. Gupta, K. Sreenivas, R. Kumar, S. Mozumdar and P.K. Patanjali 2004 Response speed of SnO2-based H2S gas sensors with CuO nanoparticles Appl. Phys. Lett. **84** 1180–1182.

[3] S.A. Patil, L.A. Patil, D.R. Patil, G.H. Jain and M.S. Wagh 2007 CuO-modified tin titanate thick film resistors as H2-gas sensors, Sens Actuators B: Chem. **123** 233–239.

[4] F. Teng, W.Q. Yao, Y.F. Zheng, Y.T. Ma, Y. Teng, T.G. Xu, S.H. Liang and Y.F. Zhu 2008 Synthesis of flower-like CuO nanostructures as a sensitive sensor for catalysis, Sens Actuators B: Chem. **134** 761–768.

[5] J.Y. Xiang, J.P. Tu, X.H. Huang and Y.Z. Yang 2008 A comparison of anodically grown CuO nanotube film and Cu2O film as anodes for lithium ion batteries, J. Solid State Electrochem. **12** 941–945.

[6] H.B. Wang, Q.M. Pan, H.W. Zhao, G.P. Yin and P.J. Zuo 2007 Fabrication of CuO film with network-like architectures through solution-immersion and their application in lithium ion batteries, J. Power Sources **167** 206–211.

[7] X.G. Zheng, C.N. Xu, Y. Tomokiyo, E. Tanaka, H. Yamada and Y. Soejima 2000 Observation of charge stripes in cupric oxide, Phys. Rev. Lett. **85** 5170–5173.

[8] M. Vaseem, A. Umar, Y.B. Hahn, D.H. Kim, K.S. Lee, J.S. Jang and J.S. Lee 2008 Flowershaped CuO nanostructures: structural, photocatalytic and XANES studies, Catal. Commun. **10** 11–16.

[9] M.-H. Chang, H.-S. Liu and C.Y. Tai 2011 Preparation of copper oxide nanoparticles and its application in nanofluid , Powder Technol. **207**, 378 .

[10] Y. She, Q. Zheng, L. Li, Y. Zhan, C. Chen, Y. Zheng and X.Lin 2009 Rare earth oxide modified CuO/CeO2 catalysts for the watergas shift reaction Int. J. Hydrogen Energy **34** 8929 .

[11] P.P.C. Udani, P.V.D.S. Gunawardana, H.C. Lee and D.H. Kim, 2009 "Steam reforming and oxidative steam reforming of metha-nol over CuOCeO2 catalysts Int. J. Hydrogen Energy **34** 7648 .

[12] J.L. Cao, G.S. Shao, Y. Wang, Y. Liu and Z.Y. Yuan 2008 CuO catalysts supported on attapulgite clay for low-temperature CO oxidation Catal. Commun. **9** 2555 .

[13] KAYANI, Zohra Nazir, et al. 2015 Characterization of copper oxide nanoparticles fabricated by the sol–gel method. *Journal of Electronic Materials* **44.10** 3704-3709.

[14] X.P. Gao, J.L. Bao and G.L. Pan 2004 Preparation and Electro-chemical Performance of Polycrystalline and Single Crystal-line CuO Nanorods as Anode Materials for Li Ion Battery , J. Phys. Chem. B **108** 5547 .

[15] R. Vijaya Kumar, R. Elgamiel, Y. Diamant, and A. Gedanken 2001 Sonochemical Preparation and Characterization of Nano-crystalline Copper Oxide Embedded in Poly(vinyl alcohol) and Its Effect on Crystal Growth of Copper Oxide Langmuir **17** 1406 .

[16] C.L. Carnes, J. Stipp and K.J. Klabunde 2002 Synthesis, Cha-racterization, and Adsorption Studies of Nanocrystalline Copper Oxide and Nickel Oxide Langmuir **18** 1352 .

[17] Fine, G. F., Cavanagh, L. M., Afonja, A., and Binions, R. 2010 Metal oxide semi-conductor gas sensors in environmental monitoring. *sensors*, ***10*(6)** 5469-5502.

[18] Liu, X., Hu, M., Wang, Y., Liu, J., and Qin, Y. 2016 High sensitivity NO2 sensor based on CuO/p-porous silicon heterojunction at room temperature. *Journal of Alloys and Compounds* **685** 364-369.

[19] Rao, M. P., Wu, J. J., Asiri, A. M., Anandan, S., and Ashokkumar, M. 2018 Photocatalytic properties of hierarchical CuO nanosheets synthesized by a solution phase method. *Journal of Environmental Sciences* **69** 115-124.

[20] M. Kaura, K.P. Muthea, S.K. Despande, S. Choudhury, J.B. Singh, N. Verma, S.K. Gupta, and J.V. Yakhmi 2006 Growth and branching of CuO nanowires by thermal oxidation of copper. J. Cryst. Growth **289** 670–675 .

[21] Zhang, L., Lu, W., Feng, Y., Ni, J., Lü, Y., and Shang, X. 2008 Facile synthesis of leaf-like Cu (OH) 2 and its conversion into CuO with nanopores. Acta Physico-Chimica Sinica **24(12)** 2257-2262.

[22] Ramizy, Asmiet, Abubaker S. Mohammed, Isam M. Ibrahim, and M. H. Eisa. 2020 Performance of Multi-Function Devices Fabricated from La2O3-Doped NiO Thin Films. International Journal of Nanoelectronics and Materials .

[23] Bouazizi, N., Bargougui, R., Oueslati, A., and Benslama, R. 2015 Effect of synthesis time on structural, optical and electrical properties of CuO nanoparticles synthesized by reflux condensation method. *Adv. Mater. Lett.* ***6*(2)** 158-164.

[24] Mohammed, A.S., Kafi, D.K., Ramizy, A., Abdulhadi, O.O. and Hasan, S.F., 2019. Nanocrystalline Ce-doped CdO thin films synthesis by spray pyrolysis method for solar cells applications. Journal of Ovonic Research Vol **15(1)** pp.37-42.

[25] ZHAO, J. G., et al. 2011 Hydrothermal synthesis and ferromagnetism of CuO nanosheets. *Applied surface science* **257.22** 9678-9681.

[26] Y. Cudennec, A. Lecerf 2003 The transformation of Cu(OH)2 into CuO, revisited, Solid State Sci. **5** 1471–1474.

[27] D.Kohl, 1989 Surface processes in the detection of reducing gases with SnO2-based devices.Sensors and Actuators, **18(1)** 71-113

[28] Shao, F., Hernández-Ramírez, F., Prades, J. D., Fàbrega, C., Andreu, T., & Morante, J. R. 2014 Copper (II) oxide nanowires for p-type conductometric NH3 sensing. *Applied surface science*, **311** 177-181.

[29] Sakthivel, B., and Nammalvar, G. 2019 Selective ammonia sensor based on copper oxide/reduced graphene oxide nanocomposite. *Journal of Alloys and Compounds* **788** 422-428.

[30] Soon, A., Todorova, M., Delley, B., and Stampfl, C. 2007 Thermodynamic stability and structure of copper oxide surfaces: A first-principles investigation. *Physical Review B* ***75*(12)**125420

Morphology, Structure, and Optical Properties of ZnO nanorods/Eosin-y Grown via Microwave-assisted Hydrothermal Method

A J Rahma[1, 2], H F Oleiwi[1,*], S G Khaleel[1] and M M Mutter[3]
[1] Department of physics, College of Science for Women, University of Baghdad, Baghdad, Iraq.
[2] Engineering Technical College, Middle Technical University (M.T.U.), Baghdad, Iraq.
[3] Ministry of Science and Technology.
jabbaraseel15@gmail.com, hindea_phys@csw.uobaghdad.edu.iq
;souadgk_phys@csw.uobaghdad.edu.iq, mahdimutter@yahoo.com

Abstract. Zinc Oxide (ZnO) was a highly attractive semiconductor material because it has a lot of properties such as non-toxic oxide, a friend of the environment, and stability for biological and biomedical applications. ZnO nanoparticle was deposited on the FTO substrates using sol-gel and spin-coating techniques and microwave-assisted hydrothermal methods were used to grow ZnO nanorods (ZNRs) from the seed layer. After that, the surface of ZNRs was modified using organic Eosin-y dye with different concentrations (0.5, 1, and 3) mM to enhance the surface of ZNRs. The XRD patterns of the as-synthesized ZNRs were indicated that the wurtzite hexagonal structure and the diameter of ZNRs will be increased after coated with Eosin-y and the surface becomes rough. The ZNRs/Eosin-y adsorbed more than 50% incident light and the electronic properties of ZNRs/1mM Eosin-y were studied using Photoluminescence (PL) spectrum and the results proved that 1 mM was the optimum concentration of Eosin-y dye that lead to increase the defects which could be enhanced the properties of ZNRs surface.

Keywords: ZnO nanorods, Eosin-y, Microwave Synthesis, Photoluminescence.

1. Introduction

ZnO was a very crucial semiconducting material because it has a petition bandgap (3.37)eV, II-VI compound semiconductors could be used in optoelectronic applications for short wavelength and could ensure efficient exciton emission at 25°C by high exciton binding energy (60 meV) in ZnO crystal [1]. Also, ZnO is diaphanous to visible light and has high conductivity at 25°C and it exhibits piezoelectric and pyroelectric properties [2]. ZnO has been consisting of specific growth conditions, so it has a rich sent family of nanostructures compare with all material [3]. The structure of ZnO could be zinc-blende, wurtzite, and rock salt and was hexagonal in wurtzite type that uses in many fabrications epitaxial thin film [4]. Hexagonal crystal of pure ZnO (wurtzite) has lattice parameters a = b = 0.32 nm and c = 0.520 nm in angle 120º[5], this structure built from the number of planes composed of Zn+2 and O-2 ions with the tetrahedral coordinate, configuration intermittent along the c axis [6]. These advantages of ZnO, which made it most commonly used, in addition to being non-toxic oxide, a friend of the environment, stable for biological and biomedical applications like diagnosis plus cancer remedy and imaging of tissue [7].

ZnO nanorod (ZNRs) which prepared using microwave method coated by an organic Eosin-y dye to improve the wettability of ZNRs and enhance charge transfer efficiency at ZNRs and increasing an absorbance for thin-film by penetrating from it so it enhanced this property [8]. The radiation of the

Microwave is one of a higher frequently type of electromagnetic field. The interaction between the molecules of material and the radiation of microwave would take place when the material was exposed to the radiation of microwave and it has plenty of advantages like uniform heating lead to deep energy penetration and reduction processing time, so it is a rapid synthesis method because of the energy is generated inside the material [9].

The interaction between microwaves with the material due to generate the heat inside the material but in the traditional method of heating thermal conductivity occurs gradually from the surface of the container to the bulk of the load. In conventional heating, the gradient of heating is transferred from outdoor to the surface of the material then to the indoor of material and the hardening of the container caused by elevated of temperature which caused overheat the material surface and loss to the energy so the growth not to be homogeneous at the surface and make large surface grains. In a microwave, the heat was transferred from indoor to outdoor of material which would be able to correct control of heating [10]. Basing on this, nanomaterials that prepared by microwave energy are improved the physical and chemical properties that promote by heat generated by this method [11]. The traditional heat is different from microwave oven because the molecules of material absorbed electromagnetic wave directly this leads to increase the temperature the interaction between microwave with the material due to generate the heat inside the material but in the traditional method of heating thermal conductivity occurs gradually from surface of the container to the bulk of the load. Microwave irradiation uses much less energy and quickly heated than conventional heating so it is used to produce high yields of products [12]. The heating in microwave happens when the molecules of material have a moment of a dipole [13]. It is rapid to generate inside the material, so it required a few minutes for heating. In this method, the material heated up immediately because it is not specified by the thermal conductivity of the container compared with traditional methods[14]. In addition, the heating of the microwave diminished energy consumption, the instantaneous automatically on-off for microwave oven operation leads to major energy saving[15]. On the other hand, in the conventional method the selective heating mechanism caused spending of energy in heating the atmosphere transporter, walls of the oven and other parts would not be affected while, in the microwave, the heat transferred from inside to outside of material which would be able to correct control of heating [10, 11, 16].

This study focuses on modification of the surface of ZNRs using different concentrations of Eosin-y dye and studies the influence of the modified structure on the physical and chemical properties of ZNRs that promote by heat generated by this method. To obtain a good thin film, fluorine-doped tin oxide (FTO) substrates (2×1.3) cm2 was washed using acetone and 2-propanol solution and cleaned via ultrasonic cleaner for (18) min, then dried. Sol-gel and spin-coating technique used to prepare the ZnO seed layer (nanoparticles). For sol-gel solution, equimolar of (0.2 M) zinc acetate dehydrate and diethanolamine as a stabilizing agent dissolved in ethanol in the tube containing a magnetic bar, then stirrer on (500) rpm at 100 °C for 10 min. Microwave-assisted hydrothermal with power 900W for 2min used to grow ZnO nanorod (ZNRs). Firstly, prepare the solution of growth by dissolving 0.04M of Zinc nitrate hexahydrate in deionized water with equimolar of hexamethylenetetramine and thrilled it for 2min. Then, put the samples that coating with the seed layer inside a glass tube vertically and add the growth solution on it. After that, the sample was rinsed carefully with deionized water to remove the residual aqueous solution and dried. ZNRs coated with different concentration solutions of Eosin-y (0.5, 1, and 3) mM in ethanol by spin coating technique. The absorption spectra of ZNRs coated with Eosin-y dye were investigated using Perkin Elmer Lambda 1800 UV-Vis spectrophotometer. The cross-sectional images and top view of ZnO nanorod determined using Field Emission Scanning Electron Microscope (FESEM) (model Leo-Supra 55VP, Carl Zeiss, Germany) linked with the EDX device to analytic the elemental mapping. X-ray diffraction technique (XRD) (SHIMADZU Japan) was used to determine the structural properties of ZNRs under conditions of diffractometer in the 2θ mode system with a copper target (CuKa 1.5406 Å). Photoluminescence (PL)

(Shimadzu RF-5301PC, Japan) used for measuring from (350-1000) nm at room temperature by 320nm excitation. Hall Effect measurements were used provide us the information about the quality of the semiconductor, the carrier's concentration and mobility, and the distance factor at room temperature, as a device (HMS-300 equipped with ECOPIA Co. Korea) was used to calculate the RH coefficient and to know the charge carriers.

2. Result and Discussion

Top-view and cross-section FESEM images with different scale bare for ZNRs is shown in Figure 1(a and b) and Figure 2 (a). These images explained that highly oriented hexagonal phase ZNRs grew linearly and perpendicularly from the FTO substrate surface having a thickness, diameter, and length in the range of 443.5nm 73±2 nm to 442±3 nm, respectively. Figure 1 (c and d) shows 1 mM Eosin-y covered on ZNRs uniformly and no agglomerate was formed onto the surface of ZNRs and the samples having a thickness and diameter in the range of 444.8nm and 86±2 nm, respectively. The samples showed a larger diameter and rich defect tips. And this will be analogous with the searches of L. Lim et al [1, 17].

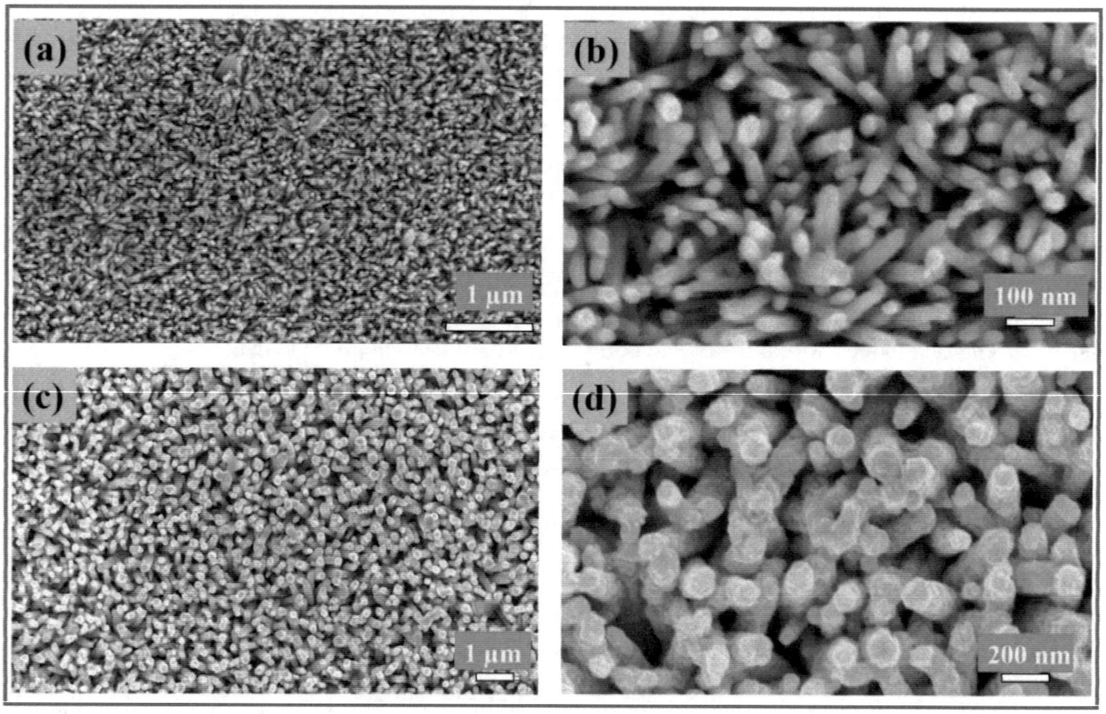

Figure (1): Top-view FESEM images with different scale bare, (a and b) for ZNRs and (c and d) for ZNRs/1mM Eosin-y prepared by microwave-assisted hydrothermal technique

Figure 2 (b) explains O and Zn elements in ZNRs by EDX analysis that linked with the FESEM device. This figure shows the element mapping of Zn and O in ZNRs which prepared by microwave-assisted hydrothermal method, this analytic shows the atomic ratio of Zn and O equals 42.29 and 44.96% respectively.

IOP Conf. Series: Materials Science and Engineering 1095 (2021) 012007 doi:10.1088/1757-899X/1095/1/012007

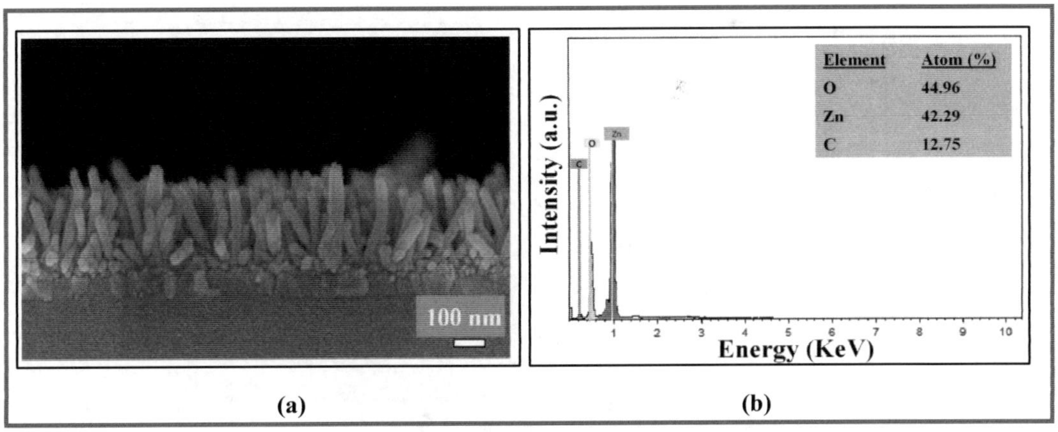

Figure (2): (a) Cross-section FESEM images of ZNRs with scale bare represents 100nm, (b) EDX spectrum of ZNRs/1mM Eosin-y

XRD patterns of ZNRs indicates the preferential orientation of the ZnO crystallites with their c-axis being perpendicular to the substrate plane, as recognized from the significantly enhanced relative peak intensity for the (002) diffraction (Figure 3). The XRD diffraction peaks of the as-synthesized ZNRs appearing at 2θ=31.75, 34.39 and 36.29 were the (100), (002) and (101) crystal planes, respectively, while, 2θ=37.59, 51.37 and 54.49 were the peaks of FTO substrate. This XRD result agrees with the database of JCPDS number 36–1451, which corresponds to the pure wurtzite hexagonal phase ZnO structure. The intensities of the XRD peaks of the as-synthesized materials indicate that they are crystalline in nature, while the broadness of the peaks confirms their small sizes. The change taking place in lattice constant (c) value was calculated with (002) basal plane that accordance with (2θ) position deviation that implies the successful incorporation of dye on the ZnO nanorod and observing the improvement of ZnO crystallinity at highly textured (002)hexagonal wurtzite structure. So the crystallinity improved when dyes of ZnO were used while maintaining the preferred orientation of ZnO in the direction (002). This a good agreement with the search [18].

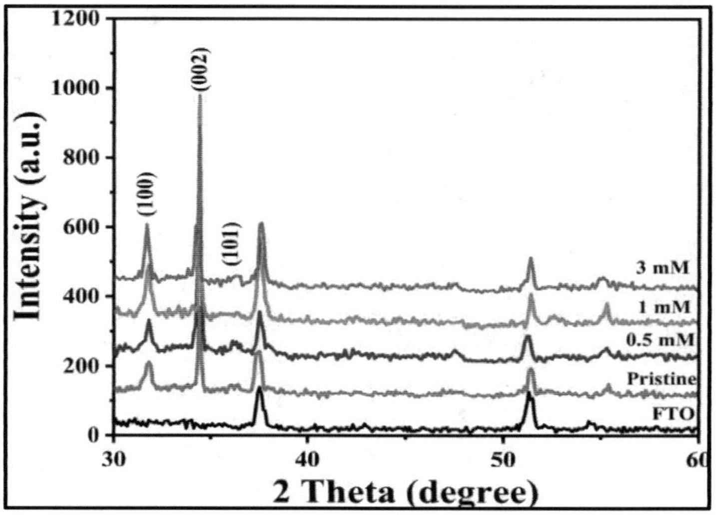

Figure (3): XRD spectra of ZNRs and ZNRs/Eosin-y films

The absorbance spectra Figure 4 (a) showing high absorption energy and abroad absorption region at the wavelength from (450-590 nm) with the same effect of concentration of Eosin-y. Figure 4 (b) explained the truncated area of absorbance spectra of ZNRs and ZNRs/Eosin-y that explained the greatest absorbance peak at (527.1 nm) wavelength which naming green region in visible light. This indicates that the absorbance peak will be increased with increasing dye concentration and this apparently related to the diameter of nanorods that calculated from FESEM results which validated that the smaller nanorods have no effect for light absorption than bigger nanorods. This indicates that the high concentrations have better absorption property, this analogous with the research M Ridwan et al.[19]. The increase of the dye concentration results in an augment of the red band absorbance. According to the Beer-Lambert law the amount of Eosin-y loaded into the hybrid films is proportional to the intensity of the absorption bands of dimer to monomer. This law is met in solutions where all dye molecules absorb photons. However, in this case the presence of dye aggregates cancel part of the expected absorption. Therefore, the intensity of the absorption spectra represents the true concentration of dye that can absorb the radiation and not the total amount of dye in the film.

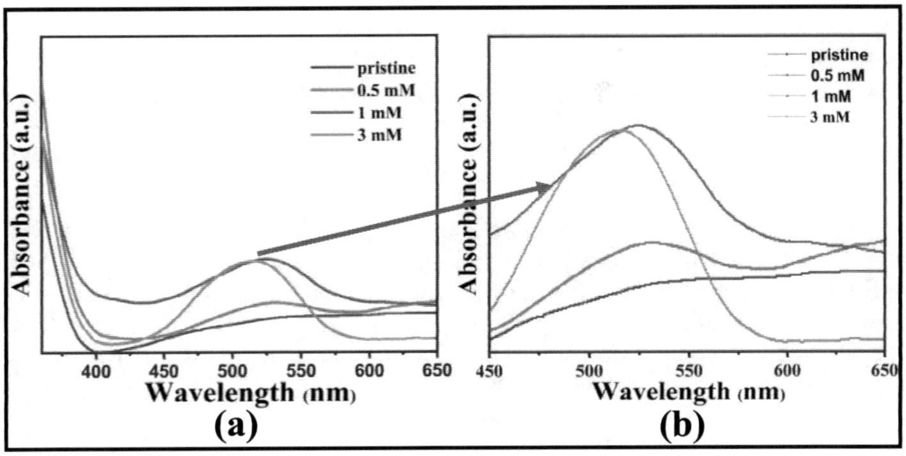

Figure (4): (a) Absorbance spectra of ZNRs and ZNRs/Eosin-y samples,(b) truncated area of absorbance spectra of ZNRs and ZNRs/Eosin-y

Figure (5) represents the transmission spectra of ZNRs which proved the behavior of the absorption spectra for the best dye concentration. However, the lower transmission obtained at (1mM) dye concentration. There was a study that proved if the thickness of the film was larger than 400nm, so the transmission of the incident light will be less than 0.1% and the absorbance will be 99.9% of incident light, the results will be agreed perfectly with the research [20].

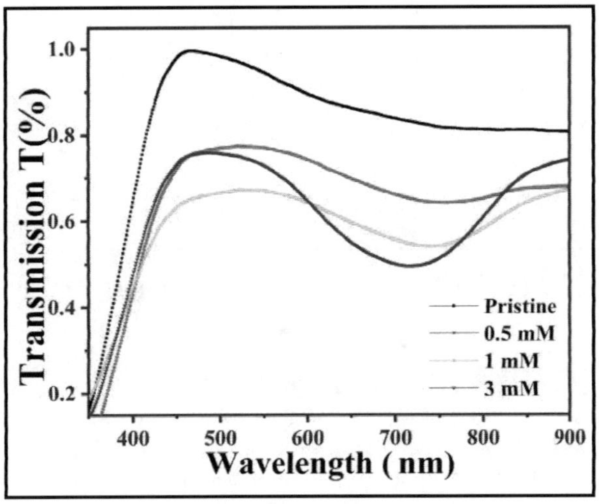

Figure (5): Transmission spectra of ZNRs and ZNRs/Eosin-y

Energy band gap (Eg) could be evaluated from optical transition by optical absorption coefficient (α) for the transition of direct electron and plotting the energy of photon (hυ) against (αhυ)2 then induction the linear accessories of curves at the energy axis and this noticed in Figure (6). In order to determine the wavelength of light necessary for the excitation of the ZNRs coated with different Eosin-y, Eg was calculated from absorption data using Schuster-Kubelka-Munk's relational expression (Equation 1).

$$(h\nu\alpha)^{\frac{1}{n}} = A(h\nu - E_g) \text{ ----- } (1)$$

Where h is Planks constant, υ is frequency of vibration, α is absorption coefficient, and A is proportionality constant. Since the direct allowed sample transition is used in this experiment, the near-UV peak and the broad visible bands are ascribed to the band-edge emission and the deep-level defects in ZnO crystal such as vacancies and interstitials of zinc and oxygen, respectively. Broad visible emission of ZNRs at 550 nm (2.25 eV) was attributed to the transition between the electron present near the conduction band and the hole at vacancy associated with the surface defects [15].

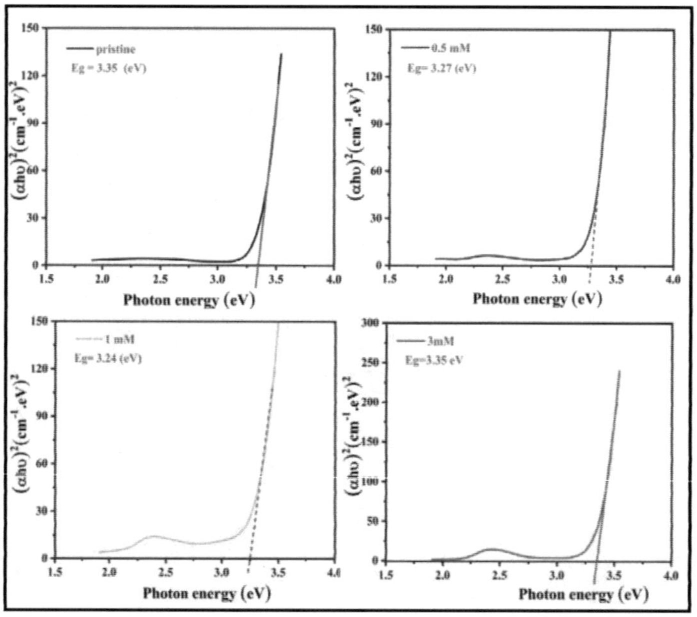

Figure (6): Tauc plot for ZNRs and ZNRs/Eosin-y samples at different dye concentrations

The Photoluminescence spectra (PL) of Eosin-y sensitized ZNRs is shown in Figure (7). The dye-adsorbed ZNRs exhibit an emission peak at 635 nm (1.952 eV) indicating a great fraction of sensitization by Eosin-y. Absorption spectra of different dye concentrations on the ZNRs appeared in the figure (7) without transmission and reflection losses. So, two peaks of absorption will appear around (680,730) nm matching to dimer and monomer of this dye [21, 22]. It explained when the concentration of dye will be increased, the absorption of this dye increased too. It is refers to more Eosin-y dye could be absorbed onto ZNRs causes increasing to absorption intensity of dimer to monomer [23-25].

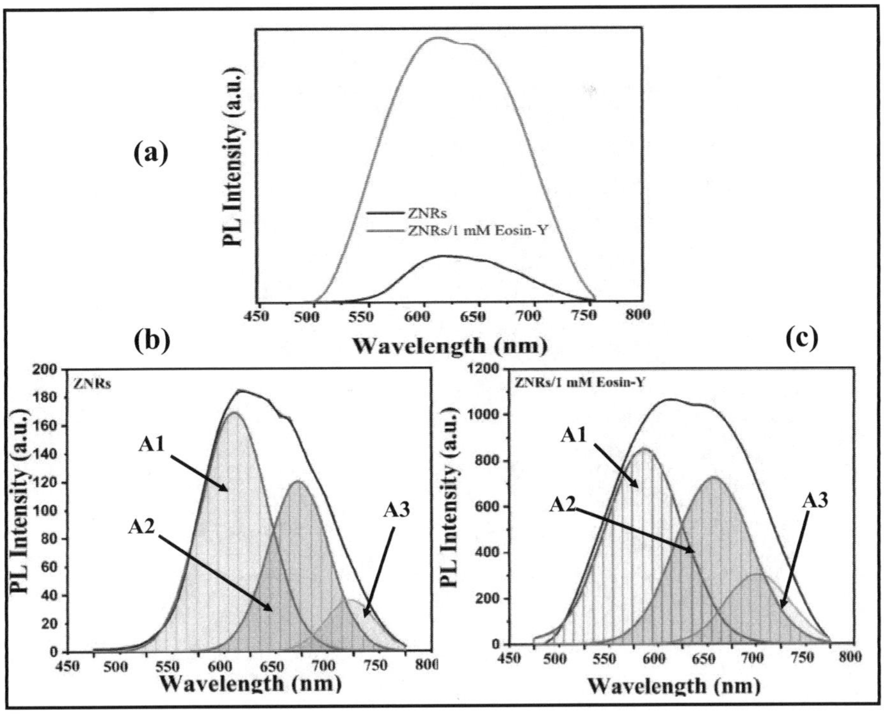

Figure (7): (a) Normalized PL spectra of ZNRs. Deconvoluted PL spectra of (b) ZNRs and (c) ZNRs/1 mM Eosin-y

Hall Effect measurement at room temperature gives information about the type of majority charge carriers, carrier's mobility, and the density of electric charge for all films. The Hall Effect was examined by the presence of a magnetic field in a vertical position recording the average change for induced Hall voltage perpendicular to the current direction passes through the film. All of these are represented in Table (1).

Table (1): The concentration and mobility carriers

	Dye Conc.	$n\times10^{19}$ (cm^{-3})	μ_H (cm^2/V. sec)	Type
Samples	0	11.325	794.18	n
	0.5	7.735	982.53	n
	1	6.250	3880.00	n
	3	10.839	610.04	n

Table (2) explained the value of the peaks which represents the kinds of defects and proved that all samples are n-type semiconductors because the donor content ratio is greater than the acceptor content ratio. And this agreement perfectly with the Hall measurements.

Table (2): The area under the curve that based on the integration peaks of ZNRs and ZNRs coated 1mM of Eosin-y and A_T represents the SUM of A_1, A_2, and A_3

Sample	ZNRs	ZNRs/Eosin-y
A_1	1374.93	7945.78
A_2	877.65	6957.55
A_3	196.94	2278.1
A_T	2449.52	17181.43
A_1/A_T	0.5613	0.4624
A_2/A_T	0.3582	0.40494
A_3/A_T	0.0803	0.13259

3. Conclusion

The objective of this research has been achieved, in which a highly uniform Eosin-y on the surface of ZNR_S that grown on FTO substrate using microwave-assisted hydrothermal specifically, the optimum concentration of this dye is 1mM which demonstrated a good properties. FESEM results proved the growth of the samples has a highly oriented and uniformly crystalline of ZNR_S on the FTO substrate, and the surface morphology has hexagonal bounded nanorod for all samples. However, the surface texture of pristine ZNR_S sample has smooth surface compare to ZNR_S/Eosin-y which become rougher. The XRD analysis demonstrated that all samples were symmetrical with ZnO wurtzite hexagonal structure and this pointing to the existence of highly crystalline of ZNR_S and ZNR_S/Eosin-y samples. Additionally, the orientation of growth for all samples is parallel to the c-axis along (0001) direction and the dominant peak oriented to the (002) Brag plane and the intensity of it will be increased with increasing the concentration of dye and no another peaks was detected in the XRD spectra despite of peaks that related to FTO substrates. The absorption light of ZNRs coated with Eosin-y was emerged in (450-590) nm in green region of visible light, this pointing to increase the absorption peak with increasing the concentration of dye and Eg of ZNR_S/Eosin-y was increased with increasing the dye concentrations. The defect of oxygen such as oxygen vacancies particularly and oxygen interstitial which could be increased when using microwave-assisted hydrothermal method and dye adding that coated nanorods samples. Hall Effect measurement showed that all films of ZNR_S and ZNR_S/Eosin-y were n-type and the mobility of carriers will be decreased with increasing the concentration of dye.

References

[1] E. L. Lim, C. C. Yap, M. Yahaya, and M. M. Salleh, "Enhancement of ZnO nanorod arrays-based inverted type hybrid organic solar cell using spin-coated Eosin-Y," *Semiconductor Science and Technology*, vol. 28, p. 045009, 2013.

[2] L. Znaidi, "Sol–gel-deposited ZnO thin films: A review," *Materials Science and Engineering: B,* vol. 174, pp. 18-30, 2010.

[3] L. Xu, Y.-L. Hu, C. Pelligra, C.-H. Chen, L. Jin, H. Huang, *et al.*, "ZnO with different morphologies synthesized by solvothermal methods for enhanced photocatalytic activity," *Chemistry of Materials,* vol. 21, pp. 2875-2885, 2009.

[4] G. A. Mohamed, A. A. El-Moiz, and M. Rashad, "Li-doping effects on the electrical properties of ZnO films prepared by the chemical-bath deposition method," *Physica B: Condensed Matter,* vol. 370, pp. 158-167, 2005.

[5] S. Pearton, D. Norton, K. Ip, Y. Heo, and T. Steiner, "Recent progress in processing and properties of ZnO," *Progress in materials science,* vol. 50, pp. 293-340, 2005.

[6] D. Rusu, G. Rusu, and D. Luca, "Structural Characteristics and Optical Properties of Thermally Oxidized Zinc Films," *Acta Physica Polonica, A.,* vol. 119, 2011.

[7] J. Li, D. Guo, X. Wang, H. Wang, H. Jiang, and B. Chen, "The photodynamic effect of different size ZnO nanoparticles on cancer cell proliferation in vitro," *Nanoscale research letters,* vol. 5, pp. 1063-1071, 2010.

[8] R. Ginting, C. C. Yap, M. Yahaya, V. Fauzia, and M. M. Salleh, "Active layer spin coating speed dependence of inverted organic solar cell based on Eosin-Y-coated ZnO nanorod arrays," in *Journal of Physics: Conference Series*, 2013, p. 012016.

[9] T. S. Tee, T. C. Hui, C. W. Yi, Y. C. Chin, A. A. Umar, G. R. Titian, *et al.*, "Microwave-assisted hydrolysis preparation of highly crystalline ZnO nanorod array for room temperature photoluminescence-based CO gas sensor," *Sensors and Actuators B: Chemical,* vol. 227, pp. 304-312, 2016.

[10] P. Yadoji, R. Peelamedu, D. Agrawal, and R. Roy, "Microwave sintering of Ni–Zn ferrites: comparison with conventional sintering," *Materials Science and Engineering: B,* vol. 98, pp. 269-278, 2003.

[11] M. Oghbaei and O. Mirzaee, "Microwave versus conventional sintering: A review of fundamentals, advantages and applications," *Journal of Alloys and Compounds,* vol. 494, pp. 175-189, 2010.

[12] J. D. Katz, "Microwave sintering of ceramics," *Annual Review of Materials Science,* vol. 22, pp. 153-170, 1992.

[13] K. Ocakoglu, S. A. Mansour, S. Yildirimcan, A. A. Al-Ghamdi, F. El-Tantawy, and F. Yakuphanoglu, "Microwave-assisted hydrothermal synthesis and characterization of ZnO nanorods," *Spectrochimica Acta Part A: Molecular and Biomolecular Spectroscopy,* vol. 148, pp. 362-368, 2015.

[14] M. Kalek, A. Ziadi, and J. Stawinski, "Microwave-assisted palladium-catalyzed cross-coupling of aryl and vinyl halides with H-phosphonate diesters," *Organic letters,* vol. 10, pp. 4637-4640, 2008.

[15] Y.-J. Jeong, C. Balamurugan, and D.-W. Lee, "Enhanced CO2 gas-sensing performance of ZnO nanopowder by La loaded during simple hydrothermal method," *Sensors and Actuators B: Chemical,* vol. 229, pp. 288-296, 2016.

[16] C. Yang, X. Su, F. Xiao, J. Jian, and J. Wang, "Gas sensing properties of CuO nanorods synthesized by a microwave-assisted hydrothermal method," *Sensors and Actuators B: Chemical,* vol. 158, pp. 299-303, 2011.

[17] M. I. A. Umar, F. Y. Naumar, M. M. Salleh, and A. A. Umar, "Hydrothermally grown of well-aligned ZnONRs: dependence of alignment ordering upon precursor concentration," *Journal of Materials Science: Materials in Electronics,* vol. 29, pp. 6892-6897, 2018.

IOP Conf. Series: Materials Science and Engineering 1095 (2021) 012007 doi:10.1088/1757-899X/1095/1/012007

[18] T. Bhat, S. Bhogale, S. Patil, S. Pisal, S. Phaltane, and P. Patil, "Synthesis and characterization of hexagonal zinc oxide nanorods for Eosin-Y dye sensitized solar cell," *Materials Today: Proceedings,* 2020.
[19] M. Ridwan, V. Fauzia, and L. Roza, "Synthesis and characterization of ZnO nanorods prepared using microwave-assisted hydrothermal method," in *IOP Conference Series: Materials Science and Engineering*, 2019, p. 012018.
[20] H. Akazawa, "Photoluminescence studies of transparent conductive ZnO films to identify their donor species," *AIP Advances,* vol. 9, p. 045202, 2019.
[21] X. Gan, X. Li, X. Gao, X. He, and F. Zhuge, "Deposition potential dependence of ZnO–Eosin Y hybrid thin films prepared by electrochemical deposition and their photoelectrochemical properties," *Materials Chemistry and Physics,* vol. 114, pp. 920-925, 2009.
[22] M. Moya, B. Marí, and M. Mollar, "Electrodeposition of hybrid ZnO/organic dye films," *physica status solidi c,* vol. 8, pp. 1744-1748, 2011.
[23] S. De, S. Das, and A. Girigoswami, "Environmental effects on the aggregation of some xanthene dyes used in lasers," *Spectrochimica Acta Part A: Molecular and Biomolecular Spectroscopy,* vol. 61, pp. 1821-1833, 2005.
[24] G. Sharma, P. Balraju, M. Kumar, and M. Roy, "Quasi solid state dye sensitized solar cells employing a polymer electrolyte and xanthene dyes," *Materials Science and Engineering: B,* vol. 162, pp. 32-39, 2009.
[25] A. Kyaw, X. Sun, C. Jiang, G. Lo, D. Zhao, and D. Kwong, "An inverted organic solar cell employing a sol-gel derived ZnO electron selective layer and thermal evaporated MoO 3 hole selective layer," *Applied Physics Letters,* vol. 93, p. 221107, 2008.

NO_2 gas sensor properties of In_2O_3-CuO Nanocomposite thin films prepared by chemical spray pyrolysis

Abd alhameed A. Hmeed[1] , Hamid S. AL-Jumaili[2]

Department of Physics, College of Education for pure Sciences,

University Of Anbar, Ramadi, Iraq .

Corresponding author : hameed.alamery1234@gmail.com, Eps.hamid.abtan@uoanbar.edu.iq

Abstract . The NO_2 gas sensor was synthesized from In_2O_3-CuO Nanocomposite thin films,deposited on glass substrate, with different mixing ratios (5, 15)Vol% of copper oxide, by chemical spray pyrolysis technique, at 400°C. The X-ray diffraction results showed that all the prepared films are polycrystalline, crystalline size in the range of (27.9 - 13.79 nm) with mixed ratio from 0 -15 Vol% of CuO respectively. From atomic force microscopy (AFM), the grain size ranged from (107.5 - 52.35 nm) , and the prepared films have average roughness of (3.13-17.7 nm) . The results of (UV-visible spectrometer) showed that the prepared films have high transmittance, and direct energy gap (3.5 -3.15) eV. The prepared films showed a high sensitivity to NO_2 with a concentration of 193ppm. For the unmixed films, the sensitivity reached (89.77%) with an equal response and recovery times of (24s), at 200°C operating temperature. The mixture thin films gave less sensitivity than unmixed indium oxide thin films .

Keywords : In_2O_3-CuO , NO_2 gas sensor , Nanocomposite , Optical and structural properties , chemical spray pyrolysis

1. Introduction

Transparent metal oxides (TCOs) have many applications in the field of air pollution monitoring and the detection of toxic and explosive gases [1, 2]. Transparent metal oxides are used to detect gases. Not only because of their superior thermal and physical stability, but also because of their ability to detect gases of very low concentrations [3]. Moreover, transparent metal oxide films are technologically important, due to their high optical transparency in the visible region, a wide energy gap, and good electrical conductivity [4]. Several attempts have been made to enhance the sensitivity and selectivity of Nanomaterials for semiconductor oxides, through surface modification, doping, and mixing of semiconducting metal oxides [5-8]. The mixing process of oxides affects the sensitivity and selectivity properties depending on the type of oxides and the type of the reacting gas. As a certain mixture gives good properties of a gas, and fewer properties of another gas, due to the change of the surface of the prepared composite. Where the type of surface , the p-n junction, the roughness of the surface, surface porosity, and some other factors. Among the family of transparent semiconductor oxides that have taken a lot of attention as gas sensors, indium oxide is n-type semiconductor. It can provide oxygen concerns, which helps to provide electrons to oxygen in the atmosphere, easily forms oxygen adsorption. Also has a wide energy gap, it ranges from (3.5 -3.75) eV, and has a cubic structure [4,9]. Copper oxide is a p-type semiconductor with a narrow energy gap ranging from (1.2 -

1.9) eV . It has a special importance recently in many photovoltaic applications due to its superior ability to absorb solar energy, heat emission, and other interesting photoelectric properties. [10, 11]. As a result of the development of the industry, the content of NO_2 gas in the atmosphere is constantly increasing , mainly resulting from combustion processes, which severely affects the human respiratory system [12]. In this work, a sensor for NO_2 gas was made from Nano thin films of indium oxide and copper oxide with different mixing ratios of copper oxide. Deposited on glass substrate, by chemical spray pyrolysis method, and the optical, structural and sensitivity properties of the prepared films toward NO_2 were studied.

2. Experimental part

The indium solution was prepared from indium chloride ($InCl_3$), and the copper solution from copper chloride ($CuCl_2.2H_2O$), by dissolving them separately in distilled water with a molarity of (0.01M). The slides were clean with detergent water and with ethanol using ultrasonic bath for ten minutes. The solutions sprayed automatically on glass substrate at 400°C using comprised air of 1 bar as a gas carrier at a period time of spraying (5 s) and (20 s) wait alternately, with a deposition rate of (2.5 ml / min).. The structural properties were studied using XRD (Shimadzu 6000PW 1020A°Target: Cu-Ka, Scan mode: continuous scan, Current :30mA, Voltage: 40KV, Wavelength 1.5406 A). AFM(AA3000). The optical properties were studied by Uv-Visible spectrometer type (sp-8001) in the wavelength range 300-1100nm. The sensitivity of the prepared films was measured, toward NO_2 gas, with different concentrations and with different operating temperature .

3. Results and discussion

3-1- Structural properties

X-ray diffraction (XRD) results showed that all the prepared thin films were polycrystalline. Figure (1) show the unmixed indium oxide thin films have several crystal peaks at(211), (222), (400), (143), (440), (622) for the diffraction angles of (21.43, 30.51, 35.37, 45.5, 50.92 and 60.56) respectively. The predominant phase is (222) with a cube structure, in accordance with the card (JCPDS 01-088-2160) and this corresponds to the researcher (Jiangwei Ma et.al) [13] and the researcher (Xin Li et. al) [14], with an average crystalline size (27.9nm) according to schereer-Deby formula . When adding 5% and 15% of the copper solution to the indium solution, the diffraction angles of indium oxide to the greater value were shifted , decreased their intensity, and also led to a decrease in the crystal size to the (21.2 nm) and (13.79 nm) respectively . At the mixing ratio (15%) of the copper solution, and in conformity with the card (JCPDS 98-002-9275) it was found that it belongs to the compound $In_2Cu_2O_5$, which has a monoclinic structure, and this is consistent with what the researcher (Ying Su) found [15].

Atomic force microscopy (AFM) tests showed that all the prepared films possessed a homogeneous surfaces. It is evident from Figure (2) that unmixed indium oxide films (0%) have average roughness equal to (3.13nm), and the root mean square root (3.84 nm) , and a grain size in the order of (107.5 nm). When adding (5%) of the copper solution, it led to decrease in the average grain size to (84.92 nm), and an increase in the average roughness of surface to (7.56 nm), and the root mean square to (8.83nm). The thin film grew vertically, and the shape of the granules became a mixture of circular and elliptical granules. When adding (15%) of the copper solution, the average grain size was decreased to (52.35nm), as well as the continuous in increase in the average of surface roughness to (17.7 nm), root mean square to (20.5nm), and the shape of the grains has become circular. We conclude from the above that the addition of the copper solution to the indium solution led to a decrease in the grain size and an increase in the average of roughness and root mean square, as well as a change in the shape of the grains from elliptical to circular shape. This is in agreement with the researcher (A H Sofi et al.) [16] and the researcher (Mahdi et al.) [17].

3-2- Optical properties

Figure (3) shows the optical transmittance of unmixed indium oxide films (0%), and the mixture films with copper oxide of two mixing ratios of copper solution (5.15) Vol% . Where the figure shows that the unmixed indium oxide thin films have a high transmittance, reaching (74%) at a wavelength (650nm).While when adding a copper solution of (5%), it decreased to (68%) at a wavelength (650 nm), and at a mixing ratio of (15%) it reached (70.5%) at the same wavelength. The reason for the decrease in optical transmittance with the increase in the mixing ratio of the copper solution is due to the high absorption of copper oxide within the visible region of the wavelength, due to the lower value of the energy gap, and this is in agreement with the researcher (Mahdi et al.) [17].

Figure (4) represents the optical energy gap values of the direct type, which obtained by drawing the graphical relationship between the energy of the photon ($h\upsilon$) and $(\alpha h\upsilon)^2$, it was equal to (3.5 eV),for unmixed films and (3.4eV) , and (3.15 eV) for the mixed films with 5% and 15% of copper solution . The reason for decreasing the energy gap value with increasing the percentage of copper solution is to anew states in the forbidden gap due to the small value of the energy gap of copper oxide.

4. Sensitivity properties

Figures (5), (7) and (8) show the change of electrical resistance with the reaction time of the gas for the prepared sensors towards NO_2 gas with a concentration of 193 ppm, and at different operating temperatures. Nitrogen dioxide (NO_2) gas is considered an oxidizing gas, just as indium oxide is of the n- type, so when the sensor thin films is exposed to the NO_2 gas, the interaction between them increases the electrical resistance of the sensor, The sensitivity was calculated, depending on the following equation:

$$S\% = \frac{R_g - R_a}{R_g} \times 100\% \quad \text{------------------(1)}$$

Where Ra is the electrical resistance of the sensor in the air, Rg is the electrical resistance of the gas sensor. Figure (5) shows the change in electrical resistance of unmixed indium oxide films with gas reaction time, at different temperatures, to give sensitivity of (45.99%), response time (16s) and recovery time (20s), at a temperature 100°C, when increasing the operating temperature to 200°C the sensitivity become (89.77%) with response time(18 s) and recovery (8s), as shown in Table (1).

In order to know the effect of the change in the concentration of NO_2 gas on the sensitivity, different concentrations were taken at a temperature of 200°C as in Figure (6), which shows the change of electrical resistance with time of the unmixed indium oxide films deposited on glass substrate . It was found that the sensor has a good sensitivity at low concentrations (10ppm) of (46.15%), and the sensitivity value increased with the increase in the concentration ratio, to (40ppm) to arrive (71.71%), at the same operating temperature. We previously concluded that the unmixed indium oxide films have good sensitivity even at low concentrations, and that increasing the NO_2 concentration above (40ppm) did not significantly affect the sensitivity values, which means that the sensor can sense the low concentrations (10ppm) with high efficiency.

When adding (5%) of the copper solution to the indium solution, it decreases the sensitivity to reaching (27.94%), at 250°C , as shown in fig.(7) and table (1). When adding (15%) of the copper solution, the measurements showed that the sensor does not work until operating temperature of 250 °C ,a value of Sensitivity at 250 °C is (12.32%), response time (12s), and recovery time (15s), as shown in Fig.(8) ,and Table (1), The add a copper solution mean copper oxide, which is p--type, and thus will form in the films p-n junction between the two oxide, and also adding copper oxide to indium oxide transforms the conductivity from n-type to p-type [18]. Therefore, the reaction with the NO_2 gas will be in two states of oxidation and reduction, which led to a decrease in sensitivity when mixed, and the reason is due to the interaction between copper oxide and NO_2 gas, and this is in agreement with the researcher (DM Chethana et.al) [19]. In the previous paper {20} under publishing , the sensitivity of the composite In_2O_3-CuO to NH_3 was greater than that of NO_2, for unmixed oxide films .This means the selectivity of this composite is high towards different gases, and this is consistent with what the researcher found (Xiaoxin Li et.al) [21].

The sensitization mechanism for the semiconductor metal oxides devices falls under two different models: the ionic absorption model includes changes that take place in the sensor surface, including oxidation, reduction and gas adsorption reactions, while the oxygen vacancy model focuses on the interaction between gas molecules and oxygen ions adsorbed on the surface [22, 23]. The adsorbed oxygen ions appear on the surface of the sensor when exposed to air, which leads to the formation of a depletion layer and a voltage barrier as a result of the capture of electrons from the conduction band, which increases the resistance of the sensor [24]. The n-type, when the sensor is exposed to the NO_2 gas, the gas molecules interact with the oxygen adsorbed on the surface, so the gas molecules capture electrons from the surface of the sensor, which leads to an increase in the resistance of the sensor, as shown in the following equations [18]:

$$NO_{2(g)} + e^- - - - - - \rightarrow NO^-_{2(ads)} \text{-------------}(2)$$

$$NO_{2(g)} + O^-_{(ads)} + e^- - - - - \rightarrow NO^-_{2(ads)} + 2O^-_{2(asd)} \text{------}(3)$$

By observing the sensitivity values of the prepared unmixed or mixed thin films, it show that there is a certain temperature at which the sensor starts working and that the sensitivity increases with increasing the temperature to a certain extent. This means that the sensitivity of the prepared films to NO_2 gas needs activation energy for the completion of the interaction between oxygen ions and gas molecules ,we observe through the measurements, the sensor did not give any response at room temperature. We also note that the increase in the operating temperature above the certain temperature led to a decrease in sensitivity due to the a desorption of oxygen ions from the surface.

5. Conclusion

The possibility of preparing a Nano composite from the mixture of indium oxide - copper oxide by the chemical spray pyrolysis technique. From XRD and AFM results, all films were polycrystalline, and the crystalline size and grain size are decreases with increasing the ratio of copper solution . Optical tests showed that unmixed indium oxide thin films have high transmittance and decrease with increasing mixing ratio from copper solution. Also the direct optical energy gap decreases with increasing mixing ratio. The sensitivity measurements toward NO_2 showed that unmixed indium oxide films have a sensitivity higher than that of the mixture , and give a sensitivity even at low concentrations of 10ppm.

References

1. Rai, Prabhakar,Raj, Sudarsan,Ko, Kyoung-Jun,Park, Kyung-Keun,Yu, Yeon-Tae, Synthesis of flower-like ZnO microstructures for gas sensor applications. Sensors and Actuators B: Chemical, 2013. **178**: p. 107-112.

2. Flak, Dorota,Braun, Artur,Mun, Bongjin Simon,Park, Jong Bae,Parlinska-Wojtan, Magdalena,Graule, Thomas,Rekas, Mieczyslaw, Spectroscopic assessment of the role of hydrogen in surface defects, in the electronic structure and transport properties of TiO_2, ZnO and SnO_2 nanoparticles. Physical Chemistry Chemical Physics, 2013. **15**(5): p. 1417-1430.
3. Kim, Nam-Hoon,Choi, Seon-Jin,Yang, Dae-Jin,Bae, Jihyun,Park, Jongjin,Kim, Il-Doo, Highly sensitive and selective hydrogen sulfide and toluene sensors using Pd functionalized WO_3 nanofibers for potential diagnosis of halitosis and lung cancer. Sensors and Actuators B: Chemical, 2014. **193**: p. 574-581.
4. Bouhdjer, A,Attaf, A,Saidi, H,Bendjedidi, H,Benkhetta, Y,Bouhaf, I, Correlation between the structural, morphological, optical, and electrical properties of In_2O_3 thin films obtained by an ultrasonic spray CVD process. Journal of Semiconductors, 2015. **36**(8): p. 082002.
5. Fu, Tiexiang, CuS-doped CuO nanoparticles sensor for detection of H2S and NH3 at room temperature. Electrochimica Acta, 2013. **112**: p. 230-235.
6. Nayel, H.H .and H.S. AL-Jumaili, Fabrication of Highly Sensitive NH3 Sensor Based on Mixed In_2O_3–AgxO Nanostructural Thin Films deposited on porous silicon. Journal of university of Anbar for Pure science, 2019. **13**(2): p. 40-47.
7. Sun, Yongjiao,Zhao, Zhenting,Suematsu, Koichi,Li, Pengwei,Yu, Zhichao,Zhang, Wendong,Hu, Jie,Shimanoe, Kengo, Rapid and Stable Detection of Carbon Monoxide in Changing Humidity Atmospheres Using Clustered In_2O_3/CuO Nanospheres. ACS sensors, 2020. **5**(4): p. 1040-1049.
8. Mohammed, N. and A. JHS, Synthesis of ZnO-SnO_2 nano thin films on porous silicon as NH3 gas sensing performance .J-Ingineering and Application Sciences, 2019. **14**(15)-16.
9. Zhou, Jiao,Ikram, Muhammad,Rehman, Afrasiab Ur,Wang, Jing,Zhao, Yiming,Kan, Kan,Zhang, Weijun,Raziq, Fazal,Li, Li,Shi, Keying, Highly selective detection of NH_3 and H_2S using the pristine CuO and mesoporous In_2O_3@ CuO multijunctions nanofibers at room temperature. Sensors and Actuators B: Chemical, 2018 :255 .18 p. 1819-1830.
10. Xia, Yusheng,Pu, Xuxin,Liu, Jie,Liang, Jie,Liu, Pujun,Li, Xiaoqing,Yu, Xibin, CuO nanoleaves enhance the c-Si solar cell efficiency. Journal of Materials Chemistry A, 2014. **2**(19): p. 6796-6800.
11. Nikam, Arun V,Arulkashmir, Arulraj,Krishnamoorthy, Kothandam,Kulkarni, Amol A,Prasad, BLV, pH-dependent single-step rapid synthesis of CuO and Cu_2O nanoparticles from the same precursor. Crystal Growth & Design, 2014. **14**(9): p. 4329-4334.
12. Ding, Mengdi,Xie, Ning,Wang, Chong,Kou, Xueying,Zhang, Hong,Guo, Lanlan,Sun, Yanfeng,Chuai, Xiaohong,Gao, Yuan,Liu, Fengmin, Enhanced NO_2 gas sensing properties by Ag-doped hollow urchin-like In_2O_3 hierarchical nanostructures. Sensors and Actuators B: Chemical, 2017. **252**: p. 418.427-
13. Ma, J., et al., High sensitivity and ultra-low detection limit of chlorine gas sensor based on In_2O_3 nanosheets by a simple template method. Sensors and Actuators B: Chemical, 2020. **305**: p. 127456.
14. Li, X., X. Xiong, and Q. Zhang, Performance-enhancing ultraviolet photodetectors established on individual In_2O_3 nanowires via coating a CuO layer. Materials Research Express, 2017. **4**(4): p. 045018.
15. Su, C.-Y., et al., Synthesis of $Cu_2In_2O_5$ and $CuInGaO_4$ nanoparticles. Thin solid films, 2013. **531** :p. 42-48.

16. Sofi, A. and M. Shah, Structural and electrical properties of copper doped In_2O_3 nanostructures prepared by citrate gel processes. Materials Research Express, 2019. **6**(4): p. 045039.
17. Suhail, M.H., H.S. Al-Jumily, and O.G. Abdullah, Characterization and NO 2 gas sensing performance of CdO: In_2O_3 polycrystalline thin films prepared by spray pyrolysis technique. SN Applied Sciences, 2019. **1**(1): p. 69.
18. Rahmani, M.B., et al., Transition from n-to p-type of spray pyrolysis deposited Cu doped ZnO thin films for NO_2 sensing. Sensor Letters, 2009. **7**(4): p. 621-628.
19. Chethana, D., et al., Synthesis, structural, magnetic and NO_2 gas sensing property of CuO nanoparticles. Ceramics International, 2020.
20. Abd alhameed A. Hameed , Hamid S. AL-Jumaili , Preparation and Characterization of In2O3-CuO Nanocomposite Thin Films as NH3 Gas Sensor , under puplishing
21. Li, X., et al., CuO-In_2O_3 core-shell nanowire based chemical gas sensors. Journal of Nanomaterials, 2014. **2014**.
22. Ferro, R., J.A. Rodriguez, and P. Bertrand, Development and characterization of a sprayed ZnO thin film-based NO_2 sensor. physica status solidi (c), 2005. **2**(10): p. 3754-37.57
23. Su, P.-G. and S. L. Peng, Fabrication and NO_2 gas-sensing properties of reduced graphene oxide/WO_3 nanocomposite films. Talanta, 2015. **132**: p. 398-405.
24. Kumar, N., et al., Probing the highly efficient room temperature ammonia gas sensing properties of a luminescent ZnO nanowire array prepared via an AAO-assisted template route. Dalton Transactions, 2014. **43**(15): p. 5713-5720.

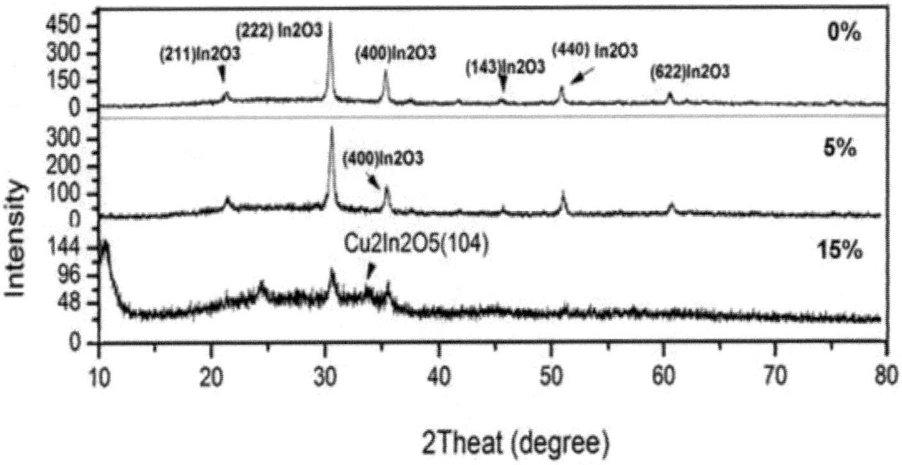

Figure (1) X-ray diffraction of the prepared films

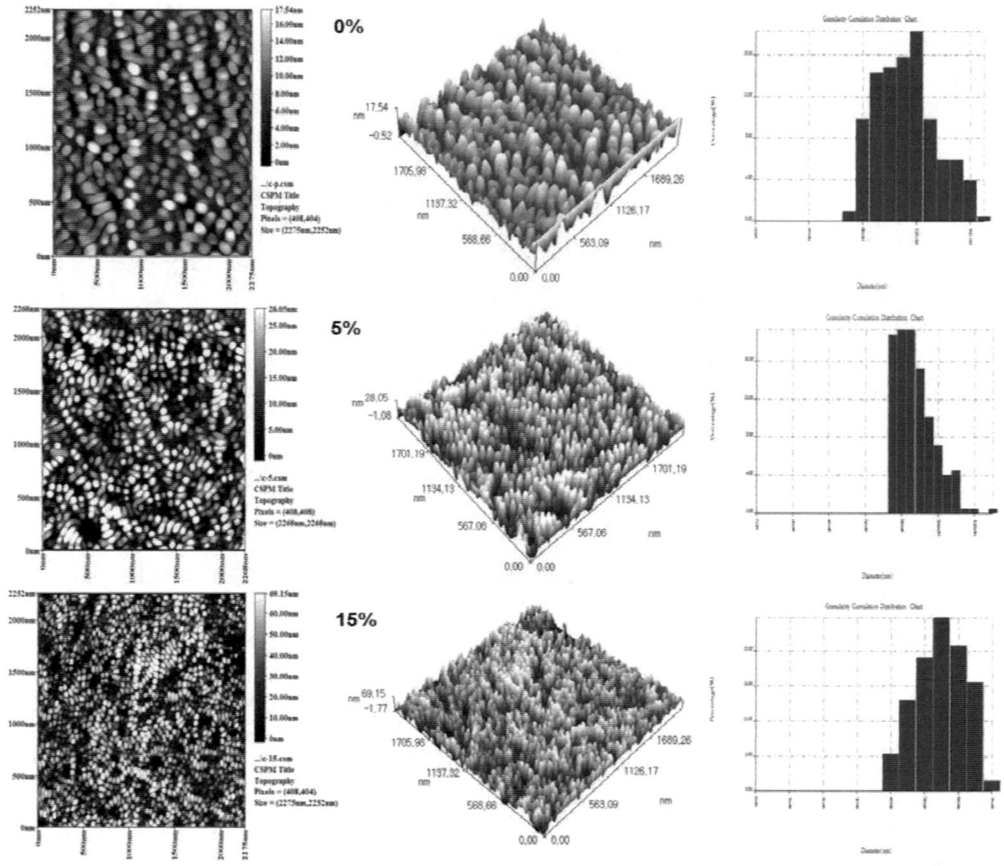

Figure (2) The AFM images of the prepared films

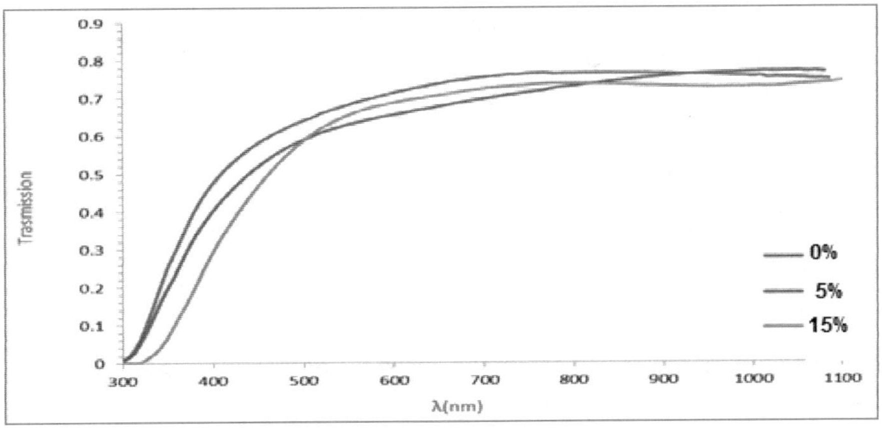

Figure (3) The transmittance values for the wavelength of the deposited films

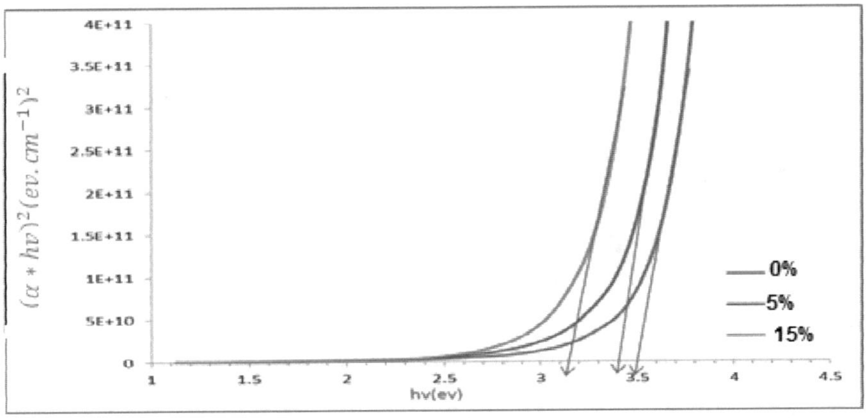

Figure (4) The optical energy gap values for the prepared films

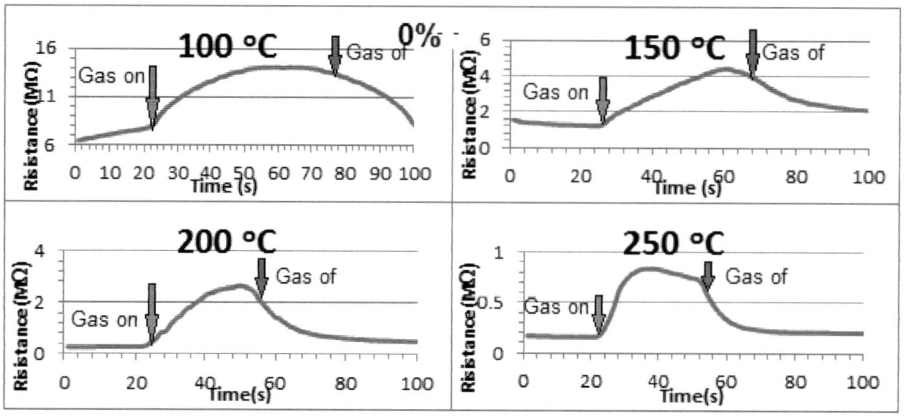

Figure (5) The change in electrical resistance with time for unmixed indium oxide films deposited on glass towards NO_2 gas at a concentration of 193 ppm .

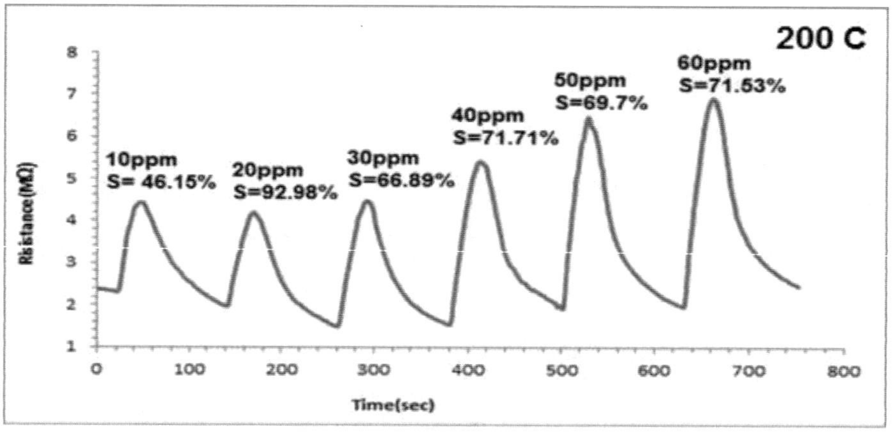

Figure (6) The change in electrical resistance with the reaction time for single indium oxide films (0%) at different NO2 concentrations.

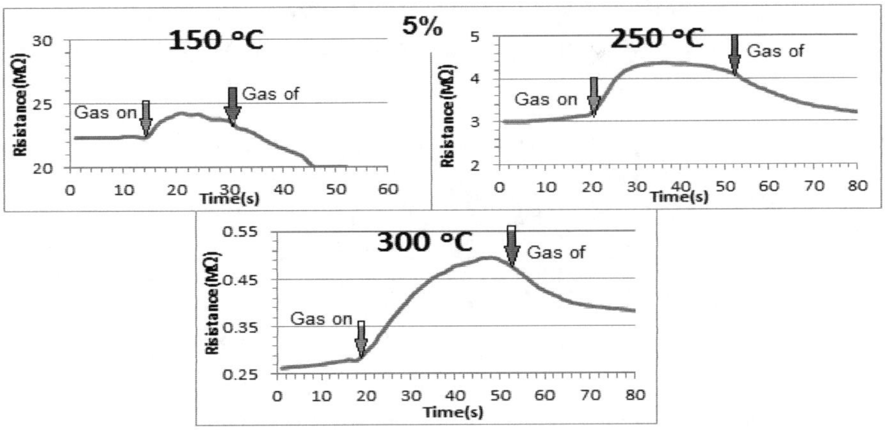

Figure (7) The change in electrical resistance with time for mixture films (5%) deposited on glass towards NO_2 at a concentration of 193 ppm.

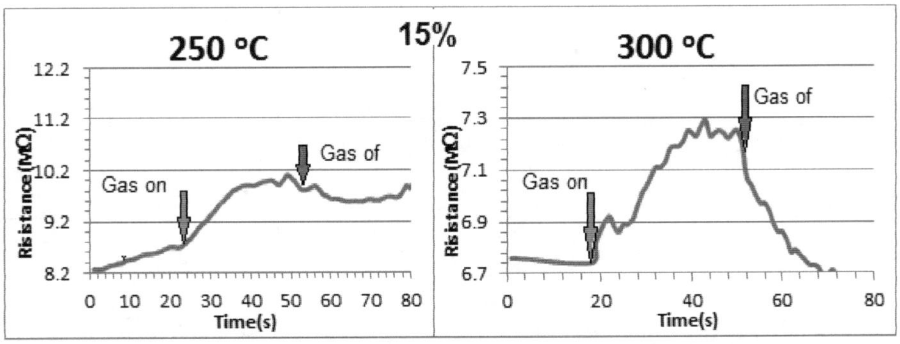

Figure (8) The change in the electrical resistance with time of the mixture films (15%) deposited on the glass towards NO_2 at a concentration of 193 ppm.

IOP Conf. Series: Materials Science and Engineering 1095 (2021) 012008 doi:10.1088/1757-899X/1095/1/012008

Table (1) The values of the sensitivity, response time and recovery time for the mixture In_2O_3-CuO films deposited on the glass substrate toward NO_2 gas at a concentration of 193ppm.

$CuCl_2$%	Sample	Operating Tem. C°	Sensitivity %	Response Time (sec)	Recovery Time (sec)
0	In_2O_3	100	45.99	16	20
		150	71.42	24	12
		200	89.77	18	8
		250	80.88	6	8
5	In_2O_3+CuO	150	6.84	8	12
		250	27.94	10	15
		300	42.94	15	10
15	In_2O_3+CuO	250	12.32	12	15
		300	5.67	17	10

Optical, Structure and Electrical Properties of $Pb_{1-x}Cd_xS$ Thin Films by Pulsed Laser Deposition (PLD) Method

Nada K Abbas [1] and Haneen S Hakeem[2]

[1]Department of Physics, College of Science for Women, University of Baghdad, Baghdad, Iraq

[2]Researcher/ Department of Physics, College of Science for Women, University of Baghdad, Baghdad, Iraq

*Corresponding authors: nadabbs@yahoo.com

Abstract. In this work, CdS, PbS, and $Pb_{1-x}Cd_xS$ nanoparticles with x ranged from 0.25 to 0.75 were synthesized using chemical co-precipitation mode. The pulsed laser deposition (PLD) method with laser system (Nd: YAG Laser system) was performed to make thin films of all samples which was normally performed under low gas pressure and under the following conditions; 300 mJ energy, 100 pulses, 6 Hz frequency and in the vacuum chamber on the substrates (10^{-2} torr). Electrical properties (D.C conductivity, Hall Effect and current-voltage measurements) of the prepared semiconductive thin films were studied using different devices. The prepared samples were characterized by UV-Vis spectroscopy (in the range 300-1100 nm) to study the optical properties, AFM and SEM to check the surface morphology (Roughness average and shape) and the particle size. XRD technique was used to determine the crystalline structure, the purity of the phase and the crystalline structure. The crystalline size averages of the nanoparticles have been found to be to 19.8, 11.0, 8.8 and 11.7 and 27.3 nm for PbS, $Pb_{0.75}Cd_{0.25}S$, $Pb_{0.5}Cd_{0.5}S$, $Pb_{0.25}Cd_{0.75}S$, and CdS respectively. The results indicate that crystalline structure of all prepared samples is cubic except CdS which shows hexagonal and cubic structure.

Keywords: CdS; Chemical co- Precipitation Deposition; PLD; PbS; $Pb_{1-x}Cd_xS$

1. Introduction:

Cadmium sulfide (CdS) is a type of chalcogenide semiconductor material with high clarity, wide and clear band gap (2.4 eV), high sensitivity of the electron, and has hexagonal and cubic structure [1]. It is useful in photochemical hydrogen production applications [2], diode sensors [3], photovoltaic cells [4], photosensor [5]. PbS semiconductor has small band gap (0.41 eV), exhibits its importance in highly efficient photovoltaic conversion devices, optical switching, and related photoelectric devices [6-8].

These properties have been correlated with the growth conditions and the nature of substrates. For these reasons, many research groups have shown a great interest in the development and study of this material by various deposition processes such as electrodeposition, spray pyrolysis, photoaccelerated chemical deposition, microwave heating and chemical bath deposition (CBD), sol–gel spin coating and pulsed-laser deposition [9-12]. Among the new and up to date techniques used for thin film deposition, pulsed laser deposition (PLD) is one of the most versatile methods to obtain layers of several materials that can be processed into a pellet target. Therefore high quality films can be obtained at a fairly low substrate temperature. Furthermore, PLD is more flexible than other conventional techniques, and it is feasible to control the thickness of films [13, 14].

Mixed thin film structures of PbS and CdS ($Pb_{1-x}Cd_xS$) have generated significant interest because they offer the advantage of tunable optical and opto-electronic properties of PbS, viz., its band gap, electrical conductivity, and structural properties (15, 16).In the current work, we report the structural and optical properties of $Pb_{1-x}Cd_xS$, where x= 0, 0.25, 0.5, 0.75,

1, thin films on glass substrate grown by PLD method at various concentration. The characterization of thin films and the optical and electrical properties were also studied.

2. Experimental:

2.1. Chemicals

The cadmium acetate [$Cd(CH_3COO)_2.2H_2O$], lead acetate [$Pb(CH_3OO)_2.2H_2O$], thiourea as a precursor of Cd^{+2}, Pb^{+2}, S^{-2} ions respectively, as well as sodium hydroxide as a stream of medium alkaline growth shall be the base materials for the preparation. The quantities and molarities of chemical compounds used in nanoparticles preparation and their procedures are previously described [17].

2.2. The Procedure

The preparation of pellets is as follows: PbS, CdS, $Pb_{0.75}Cd_{0.25}S$, $Pb_{0.5}Cd_{0.5}S$ and $Pb_{0.25}Cd_{0.75}S$ were forced for 10 minutes under pressure of 5 tons into pellet with (0.2 cm) thickness and 1 cm diameter. Hydraulic piston style SPECAC was used.
The PLD methodology was performed to make thin films of all samples. In a vacuum chamber, the method of pulsed laser deposition was normally performed under 10^{-2} torr vacuum conditions, under low gas pressure. The substrate is positioned parallel to the target surface and in front of it. There is sufficient gap between the target and the substrate, so that the holder of substrate did not block the laser beam incident. The length between the target and the substrate is 2 cm and the length is 12 cm from the target to the laser supply. PLD was employed to deposit thin films under the following conditions; 300 mJ energy, 100 pulses, 6 Hz frequency, in the vacuum chamber on the substrates.

2.3. Characterization instruments:

XRD analysis was performed using X -ray diffractometer of copper-filtered monochromatic CuKα radiation (λ=1.54 0Å), at scales 2θ (20-80), Shimadzu, Japan. Ultraviolet-Vis spectrometer (Shimadzu, Japan) has been used to measure the optical absorption obtained from solution. Atomic force microscopy (AFM), SPMAA 3000, Advanced Angestrum Inc., USA, and field emission scanning electron microscopy (FE-SEM), (SEM; Oxford instruments model SEM: S-3200N), analyzed the morphology and roughness of the surface.

3. Results and Discussion:

The PbS, CdS and $Pb_{1-x}Cd_xS$ films have been characterized by X-ray diffraction scanned from 2θ = 20 to 80 and the results illustrated in figure 1.

The observed PbS diffraction peaks are at the values of 2θ; 25.94°, 30.07°, 43.03°and 51.94° correspond to the Miller indices (111), (200), (220) and (311) respectively. The PbS findings are compatible with findings contained in the table 1, which was indexed inside a cubic structure and checked using a Regular Card (JCPDS 96-901- 3403) [18, 19]. The peaks this has not been found to suit metallic Pb or its oxide. CdS diffraction peaks found are 2θ; 25.16°, 25.59°, 28.30°, 44.30° and 52.0° be aligned with Miller indices as shown in table 1 of the standard JCPDS number [86-900-8863] [95-99-96-900-8840] and is indexed inside a cubic and hexagonal structure [20, 21]. The peaks observed of diffraction are 2θ values of $Pb_{0.75}Cd_{0.25}S$, $Pb_{0.5}Cd_{0.5}S$, and $Pb_{0.25}Cd_{0.75}S$, at Miller Indices (121), (200) and (220), which is compatible with Isaac Nkrumah et.al [22], and S. Rajathy et.al [23]. These peaks of diffraction are 2θ values of 26.05°, 30.34°, 43.36°, 26.38° and 30.71°. Thus, with Cd^{+2} taking more and more of the Pb^{+2}space, the internal strain will increase and the PbCdS solid crystal structure will become unstable. The ionic range of Cd^{+2} is even smaller than that of Pb^{+2}. This indicates very specifically that the Cd^{+2} ions are quite resolved in the PbS particle. When concentration increases, leading to the decrease in grain size, the diffraction peaks are somewhat broader. In the meantime, doping-inducing structural disorder decreases the intensity of diffraction peaks.

The calculation by scherrer equation shows that the crystalline size of PbS, $Pb_{0.75}Cd_{0.25}S$, $Pb_{0.5}Cd_{0.5}S$, $Pb_{0.25}Cd_{0.75}$, $Pb_{0.25}Cd_{0.75}S$ and CdS are averaged to 19.8, 11.0, 8.8 and 11.7 and 27.3 nm respectively.

In this work, the AFM is utilized to carry out surface morphology of the prepared thin films deposited on the glass substrates. Figure 2 shows smooth and consistent adhering films around the whole surface of the glass suspension without any voids, pinholes or crevices and with significant quantities of grains. The size of grain of thin films is several tens nanometers that demonstrate the crystalline quality of the films. These results of CdS and PbS size of grain agree with the results of Nabeel A. Bakr [24], and Abdul-Majeed E. Ibrahim et.al [19]. Table 2 indicates the typical roughness and grain size of films.

SEM is used in the surface morphology research. From the figures below, figure 3, which are display nanoparticles of various diameters with spherical shape and show uniformly homogeneous surface. This result confirms the results obtained in XRD and AFM measurements. Nucleation and coalescence is the product of the difference in grain size and shape. Generally, coalescence and nucleation affect the development of grains by shifting grain boundaries [23].

The optical transmittance spectra of the $Pb_{1-x}Cd_xS$ thin films prepared at pulsed Laser Deposition are seen in figure 4. These films were calculated to transmit on average in observable wavelength areas to PbS, $Pb_{0.75}Cd_{0.25}S$, $Pb_{0.5}Cd_{0.5}S$.$Pb_{0.25}Cd_{0.75}S$ and CdS for 17.50, 23.52, 31.60, 49.51and 80.06 respectively. This results are compatible with Abdul-Majeed E. Ibrahim et.al [19] S. Rajathi et.al [23], and Sachin H. Dhawankar et.al [25] results. With the increase in CdS concentration, transmission increases. The transmission increase can be clarified by the rise in reflection and absorption. This variation (increase and decrease) in transparency relates to the structural characteristics of film, and changes in transmission are understood to depend on the characteristics of films [26].

The $Pb_{1-x}Cd_xS$ thin film absorption spectrum from 350 nm to 900 nm, figure 5, shows different concentrations. A red shift is found in the absorption edge to lower band gap, table 3 . The energy level of dopants is under the edge of the conduction band (CBE) and over $Pb_{1-x}Cd_xS$ valence band (VBE). This energy gap leads to a red transformation in the band gap and visible light absorption through charge transfers between a doping 3d electron and $Pb_{1-x}Cd_xS$ (CB or VB) or apply a power-level crystal field transition [26]. The PbS and CdS film has also been changed to dope visible illumination with a red shift in the absorption layer. Increased absorption in the visible may be due to the change from a charge transfer, which can be defined as an electron excitation from the orbital d of metal ions to $Pb_{1-x}Cd_xS$ [27]. The graph shows that the lead absorbance is greater than the CdS. The permeability is the opposite, which is known as cadmium permeability and lead absorbent.

The absorption coefficient α from the high absorption region is found for the films by equation 1 [28].

$I=I_{\circ} \exp{-\alpha t}$ ------- (1)

If I_o is the light strength incident on the sample, α shall be the absorption coefficient, and t shall have a sample thickness. The dependency of the absorption coefficient on the deposited $Pb_{1-x}Cd_xS$ films analyzed in this study with wavelengths and at various concentrations. $Pb_{0.5}Cd_{0.5}S$ and $Pb_{0.75}Cd_{0.25}S$ is a large coefficient and CdS a low coefficient of absorption as opposed to the transmittance. The absorption coefficient values obtained of PbS and CdS are in agreement with values published by Yasmeen[29]), Srinivasan [30] and Sachin H. Dhawankar [25].

The optical band gap (Eg) was known as a statistical transfer from the surface of the valence to the conductive band [27]. Figure (6) explains the plot of (αhv) based on incident radiation energy. The thin films energy differences are rising with the rise in CdS concentration. The results of energy gap obtained are in agreement with the results found by Sachin H. Dhawankar et.al[25], Isaac Nkrumah et.al [22], F. Taghizadeh Chari [20], and Ashwini B. Rohom et.al [31].

The Hall Effect of the $Pb_{1-x}Cd_XS$ film at various concentrations is utilized to calculate the electrical mobility, carrier concentration, electrical conductivity, Hall coefficient and type of carriers. Table (4) displays that Hall coefficient (P-type) of all samples is a positive and the CdS is show the largest carrier concentration, the increase of charging carrier is mainly due to the decrease of the potential barrier. In comparison, the electric mobility of hall μ_H is concluded from CdS is the smallest. The decrease in mobility is related to the opposite relationship of μ_H to N_H which is characteristic of other polycrystalline thin films despite of the possible grain borders barriers. These results suggest an improvement in electric conductivity in PbS Scattering effects on the resistivity and structure depending upon; (1) decreases significantly the peak of XRD intensities and causes decreased crystallization and dislocation, imperfections, crystal defects and (2) decrease of crystallites, which improvement grain boundary and scattering that causes resistivity. The discontinuity of grain contacts may be due to low conductivity. This result almost corresponds to F. Taghizadeh Chari et.al[20], A.M. P´erez Gonz ´alez et.al [32] and Mosiori et.al [33].

4. Conclusions:

Thin films of CdS, PbS, and $Pb_{1-x}Cd_xS$ with 0, 0.25, 0.5, 0.75, and 1) at.wt % have been prepared by chemical co-precipitation method and PLD method. The XRD analysis in thin films is shows that PbS has a cubic structure while CdS has cubic and hexagonal structure and reveals that they are free from impurities. $Pb_{0.25}cd_{0.75}S$, $Pb_{0.5}Cd_{0.5}S$ and $Pb_{0.75}Cd_{0.25}S$ have cubic structure. The 3D and 2D AFM pictures demonstrate in thin films the spherical shapes for all the samples studied and also show a non-compact surface which is not smooth and this increases the absorbance for the prepared samples. From measurement of the optical energy gap for $Pb_{1-x}Cd_xS$ nanoparticles, it has been found that: the optical transitions are direct transitions and the $Pb_{0.5}Cd_{0.5}S$ and $Pb_{0.75}Cd_{0.25}S$ are the best samples which may be used in the solar cells with high performance. SEM images of prepared samples display thin films of various diameters with spherical shape and show uniformly homogeneous surface which confirms the results of XRD and AFM. The Hall coefficient (P-type) of all sample, CdS is the largest carrier concentration, mobility of hall N_H is concluded from CdS is the smallest.

5. References:

[1] Boutebakh F Z, Batibay D, Aida M S, Ocak Y S, Attaf N 2018 Mater. Res. Express. **5** 015511.

[2] Pareek A, Dom R, Borse P H 2013 Int. J. Hydrogen Energy **38** 36–44.

[3] Saravanakumar S, Chandramohan R, Premarani R, Devadasan J J, Thirumalai J 2017 J. Mater. Sci. Mater. Electron 28 12092–12099.

[4] Ilango M S, Ramasesha S K 2018 Pramana – J. Phys. **90** 53.

[5] Shinde S K, Dhaygude H D, Chikode P P, Fulari V J 2017 J. Mater. Sci.: Mater. Electron **28** 7385–7392.

[6] Brown P R, Patrick R B, Donghun K, Richard R L 2014 ACS Nano **8 (6)** 5863- 5872.

[7] Randy J E, Matthew C B, Justin C J, Pingrong YOI, Micic A J, Nozik, A S, Alexander L E 2005 Nano Lett. 2005 **5(5)** 865- 871.

[8] Andre S, Naser A, Christopher S, Tatiana G, Pamela N 2014 Electrochimica Acta. **151** 140- 149.

[9] Nada K A, Zainab J S, Teeba H M, Lamia K A 2018 Baghdad science journal **15(2)** 198-204.

[10] Banu N N, Ravichandran K 2017 J. Mater. Sci.: Mater. Electron **28** 11584–11590.

[11] Hariech S, Aida M S, Bougdira J, Belmahi M, Medjahdi G, Gene`ve D 2018 J. Semicond. **39** 034004.

[12] Shaban M, Mustafa M, El Sayed A M 2016 Mater. Sci. Semicond. Process **56** 329–343.

[13] Pentia E, Pintilie L, Matei I, Botila T, Ozbaya E 2001 Journal of Optoelectronics and Advanced Materials **3(2)** 525 – 530.

[14] Barış A 2015 Arabian Journal for Science and Engineering **40(7)** 2085-2093.

[15] Thangavel S, Ganesan S, Chandramohan S, Sudhagar P, Soo Y, Chang-Hee H 2010 Journal of Alloys and Compounds **495** 234-237.

[16] Modaffer A, Mousa A M, Ponpon J P 2009 J. Semiconductor Tech. and Science **9(2)** 1-6.

[17] Haneen S H, Nada K A 2021Baghdad Science Journal, in press.

[18] Faris S A 2013 J. of university of anbar for pure science 7 1.

[19] Kadhim A A, Abdul-Majeed EI, Jassim M M 2017 International Journal of Physics **5(1)** 1-8.

[20] Taghizadeh C F, Fadavieslam M R 2019 Optical and Quantum Electronics **51** 377.

[21] Martínez-Landeros V H, Hernandez-Como N, Gutierrez-Heredia G, Quevedo-Lopez M A, Aguirre-Tostado F S 2019 Thin Solid Films **682** 24–28.

[22] Isaac N, Francis K A, Bright K A, Tommy I 2015 Advances in Physics **11(1)** 2954-2959.

[23] Rajathi S, Kirubavathi K, Selvaraju K 2017 Journal of Taibah University for Science **11** 1296–1305.

[24] Nabeel A B, Falah I M, Roaa J M 2017 J. Chem. Bio. Phy. Sci. Sec. C. **7(4)** 881-889.

[25] Sachin H D, Bhagwat M S 2016 International Journal of Physical Research **4(2)** 58-61.

[26] Filipovic L, Selberherr S, Mutinati G, Brunet E, Steinhauer S, Anton K.et al. 2013 Engineering Letters **21(4)** 224-240.

[27] Ziad T A, Mohammed A A 2012. Preparation of ZnO Thin Films by Pulsed Laser Deposition. MSc Thesis, University of Al-Nahrain, Baghdad .

[28] Koao L F 2013 Material properties of semiconducting nanostructures synthesized using the chemical bath deposition method. PhD thesis, Department of Physics, University of the Free State, (QwaQwa Campus), Republic of South Africa .

[29] Yasmeen Z D, Suad M K, Ayad Z M 2015 Eng. & Tech. Journal **33(9)** Part (B) scientific; 1723-1730.

[30] Sanjeev K 2015 Nano Research & Applications **1(1)** 4.

[31] Ashwini B R, Priyanka U L, Priya R J, Ganesh R B, Nandu B C 2017 J Mater Sci: Mater Electron **28** 17107–17113.

[32] P´erez Gonz ´ alez A M, Valeriano I A 2008 Revista Mexicana de Ficica. **54(2)** 112–117.

[33] Mosiori C O, Njoroge W N, Okumu J 2014 International Journal of Advanced Research in Physical Science (IJARPS) **1(1)** 25-32.

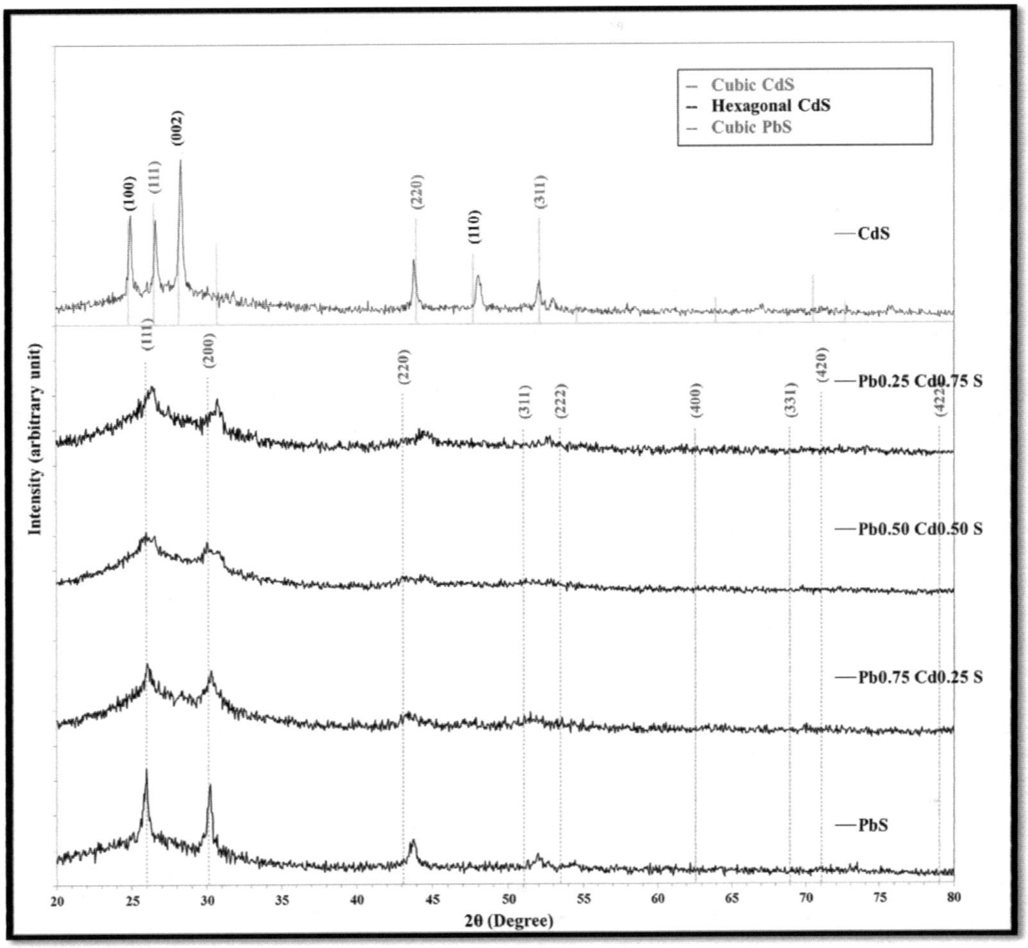

Figure 1. XRD patterns of $Pb_{1-x}Cd_xS$ thin films prepared with value of x = (0, 0.25, 0.5, 0.75and 1) wt%.

IOP Conf. Series: Materials Science and Engineering 1095 (2021) 012009 doi:10.1088/1757-899X/1095/1/012009

Table 1. The structural parameters of $Pb_{1-x}Cd_xS$ thin films.

Sample	2θ (Deg.)	FWHM (Deg.)	d_{hkl} Exp.(Å)	C.S (nm)	d_{hkl} Std.(Å)	hkl	Phase	card No.
PbS	26.0216	0.3605	3.4215	22.6	3.4246	(111)	Cub.PbS	96-901-3403
	30.2043	0.3966	2.9565	20.8	2.9657	(200)	Cub.PbS	96-901-3403
	43.7620	0.3548	2.0669	24.1	2.0971	(220)	Cub.PbS	96-901-3403
	51.9471	0.7571	1.7589	11.7	1.7884	(311)	Cub.PbS	96-901-3403
$Pb_{0.75}Cd_{0.25}S$	26.0553	0.6130	3.4172	13.3	3.4246	(111)	Cub.PbS	96-901-3403
	30.3462	0.8654	2.9430	9.5	2.9657	(200)	Cub.PbS	96-901-3403
	43.3630	0.8293	2.0850	10.3	2.0971	(220)	Cub.PbS	96-901-3403
$Pb_{0.50}Cd_{0.50}S$	26.0938	0.9096	3.4122	9.0	3.4246	(111)	Cub.PbS	96-901-3403
	30.3486	0.9538	2.9428	8.6	2.9657	(200)	Cub.PbS	96-901-3403
$Pb_{0.25}Cd_{0.75}S$	26.3904	0.7211	3.3745	11.3	3.4246	(111)	Cub.PbS	96-901-3403
	30.7173	0.6851	2.9083	12.0	2.9657	(200)	Cub.PbS	96-901-3403
CdS	25.0120	0.2524	3.5573	32.2	3.5808	(100)	Hex.CdS	96-900-8863
	26.6707	0.3246	3.3397	25.2	3.3590	(111)	Cub.CdS	96-900-8840
	28.3654	0.3245	3.1439	25.3	3.3745	(002)	Hex.CdS	96-900-8863
	43.8702	0.2884	2.0621	29.7	2.0570	(220)	Cub.CdS	96-900-8840
	48.0529	0.3605	1.8919	24.1	2.0674	(110)	Hex.CdS	96-900-8863
	52.0913	0.3245	1.7543	27.2	1.7542	(311)	Cub.CdS	96-900-8840

Table 2. Grain size and average roughness of the $Pb_{1-x}Cd_xS$ thin films.

Sample	Grain size(nm)	Roughness average(nm)
PbS	83.80	11.2
$Pb_{0.75}Cd_{0.25}S$	83.12	5.38
$Pb_{0.5}Cd_{0.5}S$	77.42	5
$Pb_{0.25}Cd_{0.75}S$	72.51	5.09
CdS	53.95	4.39

PbS

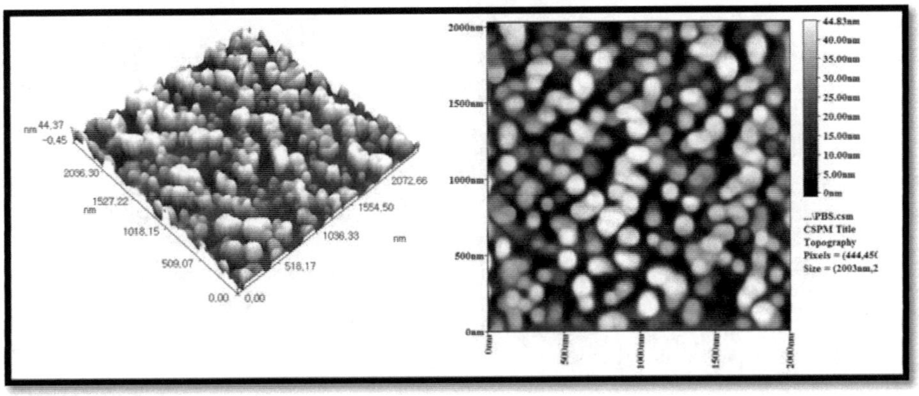

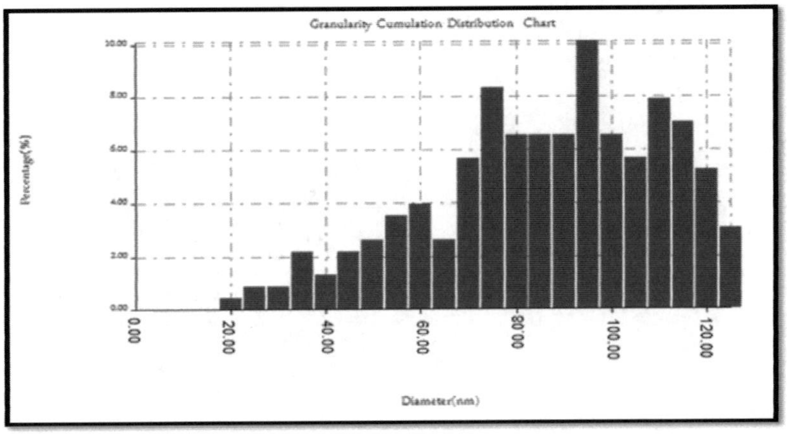

Pb$_{0.75}$Cd$_{0.25}$S

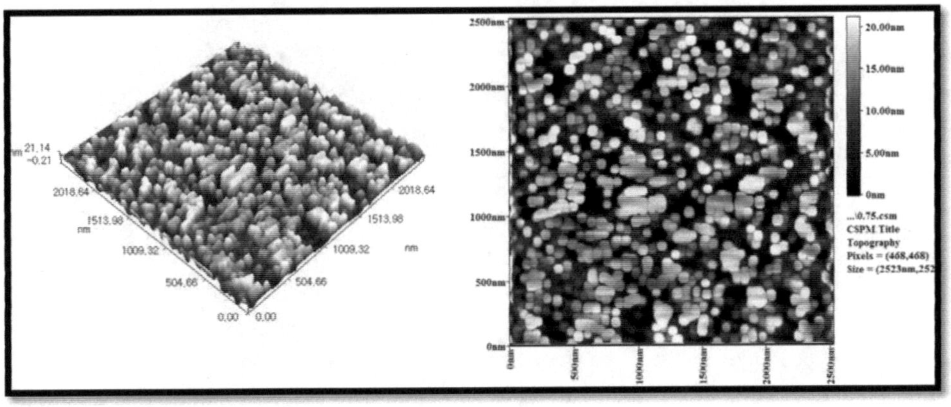

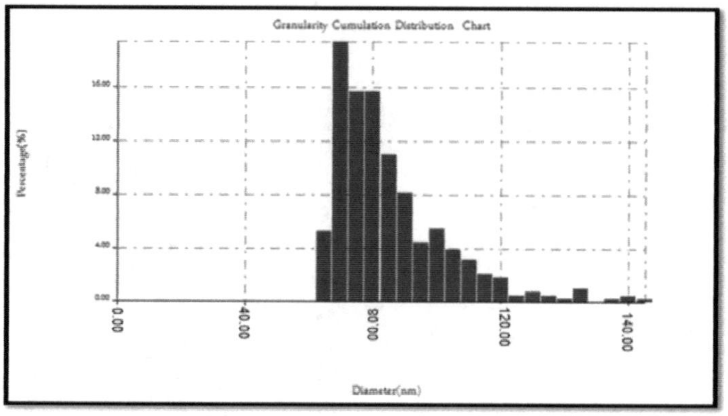

$Pb_{0.5}Cd_{0.5}S$

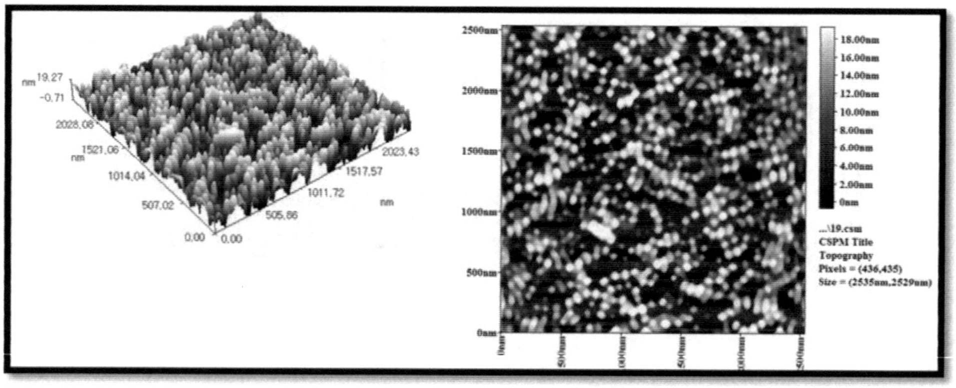

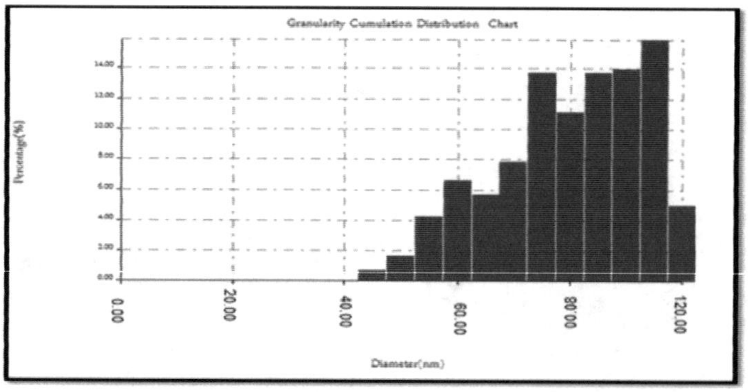

$Pb_{0.25}Cd_{0.75}S$

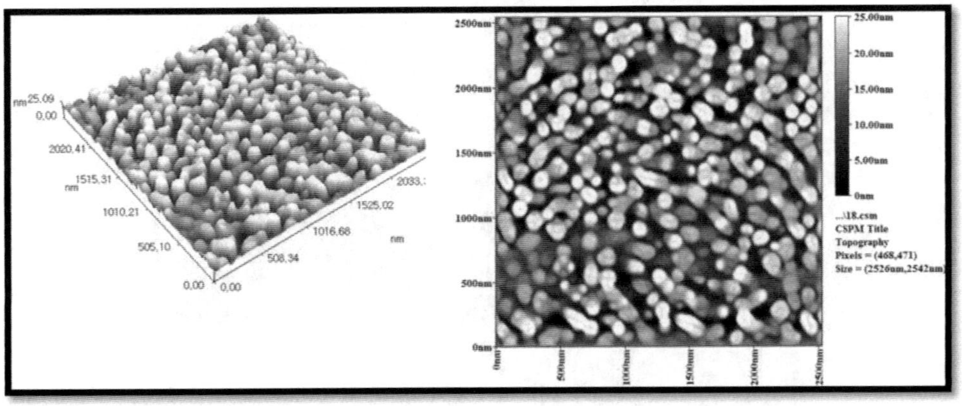
25.00nm
20.00nm
15.00nm
10.00nm
5.00nm
0nm
CSPM Title
Topography
Pixels = (468,471)
Size = (2526nm,2542nm)

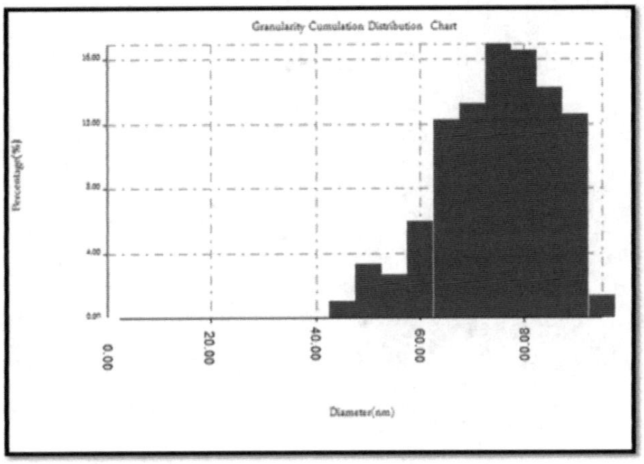
Granularity Cumulation Distribution Chart
Percentage(%)
Diameter(nm)
0.00
20.00
40.00
60.00
80.00

CdS

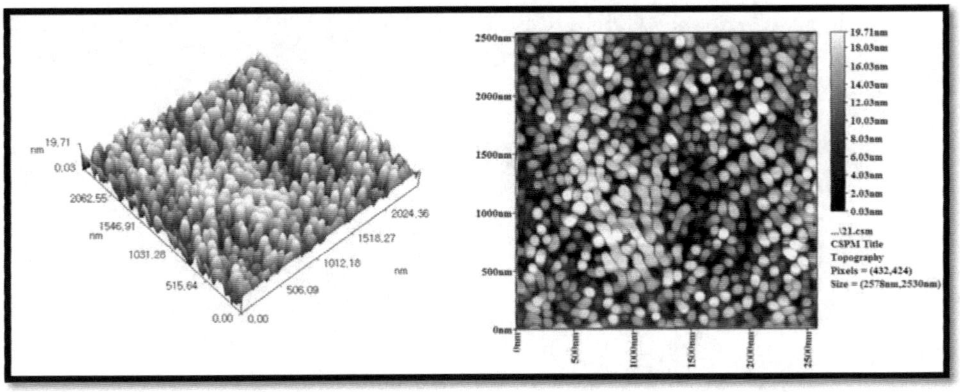
19.71nm
0.03nm
CSPM Title
Topography
Pixels = (432,424)
Size = (2578nm,2530nm)

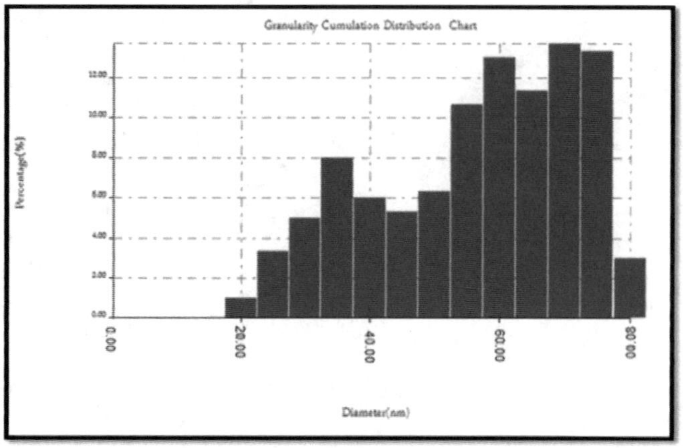

Figure 2. AFM images in 2D, 3D and grain size distribution of the $Pb_{1-x}Cd_xS$ thin films.

Figure 3. SEM images of the $Pb_{1-x}Cd_xS$ thin films.

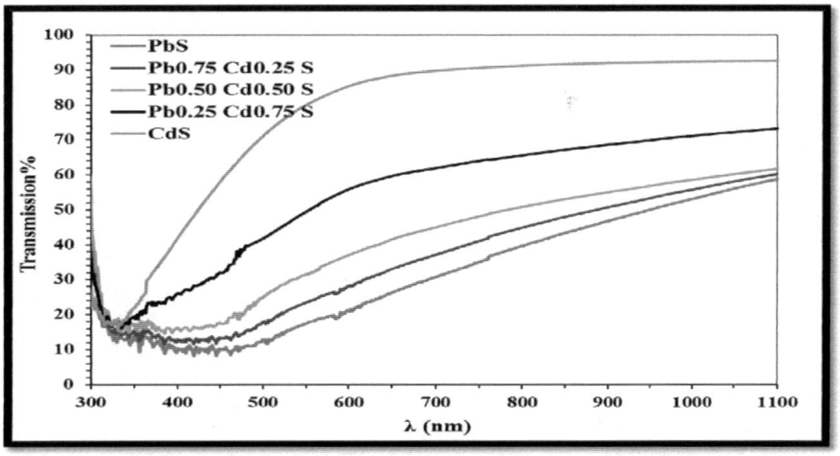

Figure 4. Optical transmission for the $Pb_{1-x}Cd_xS$ thin films.

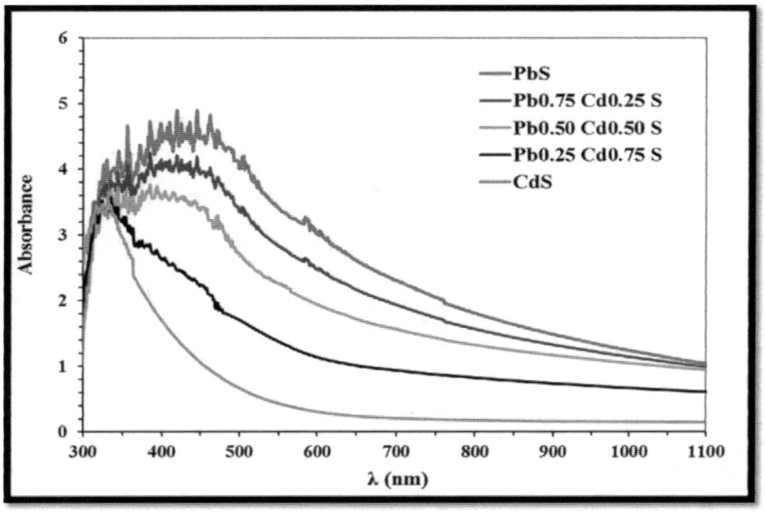

Figure 5. Absorption spectra of $Pb_{1-x}Cd_xS$ thin films.

IOP Conf. Series: Materials Science and Engineering 1095 (2021) 012009 doi:10.1088/1757-899X/1095/1/012009

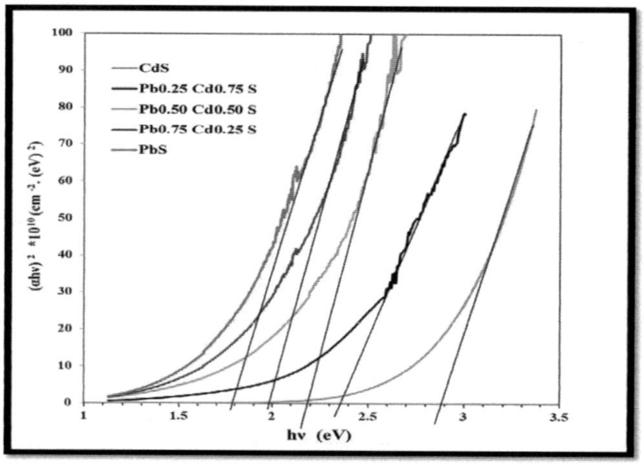

Figure 6. A plots of $(\alpha h\nu)^2$ verses photon energy (hν) of $Pb_{1-x}Cd_xS$ thin films.

Table 3. List of the optical energy gaps of $Pb_{1-x}Cd_xS$ thin films.

Sample	Eg(eV)
PbS	1.80
$Pb_{0.75}Cd_{0.25}S$	2.00
$Pb_{0.5}Cd_{0.5}S$	2.20
$Pb_{0.25}Cd_{0.75}S$	2.35
CdS	2.90

Table 4. Hall effect parameters of $Pb_{1-x}Cd_xS$ thin films with value of x=(0, 0.25, 0.5, 0.75and 1) at. wt. %.

Samples	N_H (cm)$^{-3}$	R_H (cm^{-3}C^{-1})	σ (Ω.cm)-1	μ_H (cm^{2}V^{-1}s^{-1})	Type
PbS	3.18E+14	1.97E+04	1.90E+01	3.74E+05	p
$Pb_{0.75}Cd_{0.25}S$	1.58E+14	3.95E+04	4.77E-01	1.88E+04	p
$Pb_{0.5}Cd_{0.5}S$	1.39E+14	4.50E+04	2.41E-01	1.09E+04	p
$Pb_{0.25}Cd_{0.75}S$	1.54E+14	4.05E+04	1.62E-01	6.54E+03	p
CdS	5.06E+15	1.23E+03	5.80E+00	7.15E+03	p

Some Estimates to Best Approximation of Functions by Fourier-Jacobi Differential Operators

Ehsan M Hameed[1*], Habeeb A Aal-Rkhais[1] and Salam J Majeed[1]

[1] Department of Mathematics, College of Computer Science and Mathematics, University of Thi-Qar. Iraq.

*E-mail: ehsan.m@utq.edu.iq

Abstract. In this paper, an extension of the idea of the best approximation in the Hölder spaces with respect to Fourier-Jacobi operators by moduli of smoothness is studied. A special form of the moduli of smoothness is considered to get a strong convergence. Further, advanced approaches of approximation and some direct and inverse results are proved. Moreover, the Jackson-type estimate of functions in Hölder spaces by Jacobi transformations to algebraic polynomials with generalized de la Vallée Poussin mean are established.

Keywords: Hölder spaces, Hölder function, Fourier-Jacobi, Jacobi Differential Operators

1. Introduction

In many problems, algebraic polynomials or orthogonal polynomials satisfy as solutions of ordinary or partial differential equations. Fourier-Jacobi polynomials are significant types of these polynomials. There are widely scientific and engineering applications to the Fourier-Jacobi polynomial expansions, as for instance the resolution of the Gibbs' phenomenon and electrocardiogram data compression (see [12, 5, 19]). On the other hand, several consequences of approximation theorems in Hölder spaces by algebraic polynomials have been studied, such as in [4, 13, 19].

In this work, we study the approximate of these functions that defined in the same spaces. Therefore, to analyse the harmonic of Jacobi we use a significant technique that deals with the approximation methods in periodic Hölder spaces, see [16]. Also, very important application for moduli of smoothness and the approximation of functions in Hölder spaces to estimate the behavior of asymptotic interfaces to partial differential equations in irregular domains has been considered in [1, 2].

The main result of the current work is to develop the previous results and estimate the best approximation of functions by Jacobi differential operator and algebraic polynomials with generalized de la Vallée Poussin which means (see [15]) in the Hölder spaces. Moreover, we have got a sharp result for the approximation of functions in Hölder spaces by the concept of moduli of smoothness.

Let us give a brief outline of the contents of the paper: A brief outline of the contents of the paper follows: section (2) re-states some basic concepts which are necessary for our future work. Then, some lemmas to help us in the proof of the main results in the following section are introduced. The main results of our work are presented in section (3). First, two theorems (theorem 3.1 and theorem 3.2) are introduced to estimate the best approximation of functions in the Hölder spaces by applying Fourier-Jacobi operators for the direct and inverse cases. More importantly, our main result (Theorem 3.3) shows that the best approximation converges to zero very fast. Finally, the conclusions are presented in section (4).

2. Preliminaries

Let us define $\boldsymbol{\phi} = \boldsymbol{\phi}_{\alpha,\beta}(x) = (1-x)^{\alpha}\,(1+x)^{\beta}$ on $\boldsymbol{I} = [-1, 1]$ are the Jacobi weighted functions where $\boldsymbol{\alpha}, \boldsymbol{\beta} \in (-1, +\infty)$. Also, we consider the space of functions on $\boldsymbol{I}$ which is denoted by

$$L_{\phi,p}(I) = L_{\phi,p}^{\alpha,\beta}(I) = L_p(I; \phi_{\alpha,\beta}),\ \ 1 \le p \le \infty,$$

with norm

$$\|f\|_{\phi,p} = \left(\int_I |f(x)|^p\, \phi(x)\, dx\right)^{\frac{1}{p}}.$$

When $\boldsymbol{\alpha} = \boldsymbol{\beta} = \mathbf{0}$, we have $\boldsymbol{L_p(I)} = \boldsymbol{L_p(I\,; \phi_{0,0})}$, with norm $\|\boldsymbol{f}\|_{\boldsymbol{p}} = \|\boldsymbol{f}\|_{L_p(I;\phi_{0,0})}$. Moreover, the Jacobi orthonormal polynomials and their properties are considered in significant references such as [3], [11], and [18].

Let us consider the Jacobi polynomial of order $\boldsymbol{n}$, $\boldsymbol{P_n^{(\alpha,\beta)}(x), n = 0, 1, 2, \dots}$, and it can be defined in special form (see [6]). Let $\boldsymbol{p = 2}$, we can get the normalized Jacobi polynomial under the weighted weighted $\boldsymbol{L_{\phi,p}(I)}$-inner product

$$\int_I P_m^{(\alpha,\beta)} P_n^{(\alpha,\beta)} \phi(x)\, dx = \begin{cases} 0, & n \ne m \\ \mathcal{N}_n^{(\alpha,\beta)}, & n = m \end{cases}.$$

The normalized Jacobi polynomial $\boldsymbol{J_n^{(\alpha,\beta)}(x) = P_n^{(\alpha,\beta)}(x)/\mathcal{N}_n^{(\alpha,\beta)}}$, $\boldsymbol{n = 0, 1, 2, \dots}$, where $\boldsymbol{\mathcal{N}_n^{(\alpha,\beta)}}$ is the normalization constant given as $\left\|\boldsymbol{P_n^{(\alpha,\beta)}}\right\| = \boldsymbol{\mathcal{N}_n^{(\alpha,\beta)}}$. Let $\boldsymbol{f \in L_{\phi,p}(I), 1 \le p \le \infty, \alpha, \beta > -1}$, and let

$$f(x) \cong \sum_{n=0}^{\infty} \mathcal{F}_n^{(\alpha,\beta)}(f) \mathcal{N}_n^{(\alpha,\beta)}\, J_n^{(\alpha,\beta)}(x), \tag{1}$$

represents the Fourier-Jacobi series with coefficients defined as follow

$$\mathcal{F}_n^{(\alpha,\beta)}(f) = \int_I f(x) J_n^{(\alpha,\beta)}(x)\phi(x)dx, \quad k = 0,1, \dots$$

The Fourier-Jacobi expansions are approximated to the operator $\boldsymbol{A_h^{(\alpha,\beta)}, 0 < h < \delta < \pi}$, on a function $\boldsymbol{f \in L_{\phi,p}(I)}$ with expansion (1) by the form

$$A_h^{(\alpha,\beta)} f(x) \cong \sum_{n=0}^{\infty} \mathcal{F}_n^{(\alpha,\beta)}(f) \mathcal{N}_n^{(\alpha,\beta)}\, J_n^{(\alpha,\beta)}(\cos h)\, J_n^{(\alpha,\beta)}(x), \tag{2}$$

where the operator $A_h^{(\alpha,\beta)}$ is positive (see [16]). If $f \in L_{\phi,p}(I)$, and $0 < h < \delta$, then it is easy to get from (1) and (2), that

$$\left\|A_h^{(\alpha,\beta)} f\right\|_{\phi,p} \le \|f\|_{\phi,p}. \tag{3}$$

where the restriction $\alpha \ge \beta \ge -1/2$ is guaranteed to get the inequality (3) can be satisfied.

Now, let $f \in L_{\phi,p}(I)$, $1 \le p < \infty$, and $r > 0$, then we define the r-th symmetric difference for the Fourier-Jacobi differential operator $\mathrm{A}_{\mathrm{h}}^{(\alpha,\beta)}$ is

$$\Delta_{r,h}^{(\alpha,\beta)} = \left(Id - A_h^{(\alpha,\beta)}\right)^{r/2} = \sum_{i=0}^{\infty} (-1)^i \binom{\frac{r}{2}}{k} \left(A_h^{(\alpha,\beta)}\right)^i, \tag{4}$$

where Id represents the identity operator.

Let us define the modulus of smoothness of the r-th order by

$$w_r(f, \delta)_{\phi,p} = \sup_{0<h\le\delta} \left\|\Delta_{r,h}^{(\alpha,\beta)} f\right\|_{\phi,p},$$

and for $h \geq \delta$, we define that $w_r(f,h)_{\phi,p} = w_r(f,\delta)_{\phi,p}$, (see [8,17]).
Now, we consider the Hölder spaces with respect to a general Jacobi operator $A_h^{(\alpha,\beta)}$ which is denoted by $C_{\phi,p}^{r,\mu}(I)$. For any function $f \in C_{\phi,p}^{r,\mu}(I)$, $\mu > 0$ if $f \in L_{\phi,p}(I)$ and

$$\|f\|_{C_{\phi,p}^{r,\mu}(I)} = \|f\|_{\phi,p} + |f|_{C_{\phi,p}^{r,\mu}(I)}, \tag{5}$$

is finite, where $|f|_{C_{\phi,p}^{r,\mu}(I)} = \sup_{h>0} \frac{w_r(f,h)_{\phi,p}}{h^\mu}$. Moreover, assume that $0 < \mu < r < k < \infty$ and for an arbitrary constant C, then the function $f \in C_{\phi,p}^{r,\mu}(I)$ has two norms in $C_{\phi,p}^{r,\mu}(I)$ and $C_{\phi,p}^{k,\mu}(I)$ which are equivalent, i.e.,

$$C\,\|f\|_{C_{\phi,p}^{r,\mu}(I)} \leq \|f\|_{C_{\phi,p}^{r,\mu}(I)} \leq C^{-1}\,\|f\|_{C_{\phi,p}^{r,\mu}(I)}. \tag{6}$$

Consider now the modulus of smoothness for the function $f \in C_{\phi,p}^{r,\mu}(I)$, $0 < \mu \leq k$,

$$w_r^\mu(f,t)_{C_{\phi,p}^{r,\mu}(I)} = \sup_{0<h<t} \frac{w_r(f,h)_{\phi,p}}{h^\mu}. \tag{7}$$

Some properties of these spaces can be considered in [20]. In order to get a sharp estimations and strong inequalities, we consider the following moduli of smoothness

$$\Omega_{k,r}^\mu(f,\delta)_{C_{\phi,p}^{r,\mu}(I)} = \sup_{0<h<\delta} \frac{w_k\left(\Delta_{r,h}^{(\alpha,\beta)} f,\delta\right)_{\phi,p}}{h^\mu}. \tag{8}$$

Let us now to denote the set of all algebraic polynomials by Π_n with degree n. Furthermore, we can use the algebraic polynomials with degree n to define the error of the best approximation of a function f as follow:

$$E_n(f)_{\phi,p} = \inf_{P\in\Pi_{n-1}} \|f-P\|_{\phi,p}, \quad n \in \mathbb{N},$$

where $P \in \Pi_{n-1}$ is called an algebraic polynomial of the best approximation of $f \in L_{\phi,p}(I)$ if

$$\|f-P\|_{\phi,p} = E_n(f)_{\phi,p},$$

see [7] for more details.
On the other hand, the polynomial of the best approximation of $f \in C_{\phi,p}^{r,\mu}(I)$ is represented by $P \in \Pi_{n-1}$ if

$$\|f-P\|_{C_{\phi,p}^{r,\mu}(I)} = E_n(f)_{C_{\phi,p}^{r,\mu}(I)}$$

where

$$E_n(f)_{C_{\phi,p}^{r,\mu}(I)} = \inf_{P\in\Pi_{n-1}} \|f-P\|_{C_{\phi,p}^{r,\mu}(I)}, \quad n \in N.$$

Likewise, the eigenfunctions of the differential operator can be used to define the Jacobi polynomials as follow

$$\mathcal{D} = \frac{-1}{\phi^{(\alpha,\beta)}(x)} \frac{d}{dx}\left[\phi^{(\alpha+1,\beta+1)}(x)\frac{d}{dx}\right] \tag{9}$$

$$\mathcal{D}P_k^{(\alpha,\beta)} = \lambda_k\, P_k^{(\alpha,\beta)}, \quad \lambda_k = (\alpha+\beta+k+1)k.$$

When we take $r > 0$ and $f \in L_{\phi,p}(I)$, $1 \leq p \leq \infty$, then we can get a function $g \in L_{\phi,p}(I)$, such that

$$g(x) \cong \sum_{k=1}^{\infty} (\lambda_k)^{\frac{r}{2}}\, \mathcal{N}_n^{(\alpha,\beta)}\, J_k^{(\alpha,\beta)}(x),$$

and this function is called Fourier –Jacobi series. We also can get the fractional derivative of order r of the function f when we change the notation of function g by $\mathcal{D}^r f$, i.e., $g = \mathcal{D}^r f$. In addition, there are many results related to the differential operator $\mathcal{D}^r$ which are formulated to obtain approximation in $L_{\phi,p}(I)$ in terms of the generalized $\mathcal{K}_r$-functional

$$\mathcal{K}_r(f,h)_{\phi,p} = \inf_{s}\{\|f-s\|_{\phi,p} + h^r\|\mathcal{L}^r s\|_{\phi,p}\},$$

see [7] for more details.

Further, the moduli of smoothness and $\mathcal{K}_r$-functional have a natural connection that can be explained as follow:

For $f \in L_{\phi,p}(I), 1 \le p \le \infty$, and $r > 0$, then

$$\mathcal{K}_r(f,h)_{\phi,p} \asymp w_r(f,h)_{\phi,p} \quad 0 < h < \delta. \tag{10}$$

And for $P \in \Pi_{n-1}$,

$$h^r \|\mathcal{L}^r P\|_{\phi,p} \asymp w_r(P,h)_{\phi,p}. \tag{11}$$

Let us consider the auxiliary results that we will identify in the proof of the main results in section 3. The following two lemmas are presented in [7, 14]:

Lemma 2.1. Let $f \in L_{\phi,p}(I), 1 < p \le \infty, ,0 < \mu < r$ and $k > 0$ for algebraic polynomial $P_n \in \Pi_{n-1}$ then there exists a constant C such that

$$\|f - P_n\|_{\phi,p} \le C w_{k+r}(f,n^{-1})_{\phi,p}, \quad n \in \mathbb{N}.$$

Lemma 2.2. Let $f \in L_{\phi,p}(I),\ 1 \le p \le \infty, n \in N, 0 < h < n^{-1}$, and $r > 0$ then for any algebraic polynomial $P_n \in \Pi_{n-1}$,

$$\left\|\Delta_{r,h}^{(\alpha,\beta)} P_n\right\|_{\phi,p} \le C\, w_r(f,h)_{\phi,p},$$

where C is a constant which does not depend on f, h and P_n.

Lemma 2.3. Let $f \in L_{\phi,p}(I),\ 1 \le p \le \infty$ and $0 < h < \delta$ then for any algebraic polynomial $P \in \Pi_{n-1}$,

$$w_r(f-P,h)_{\phi,p} \le w_r(f,h)_{\phi,p} + w_r(P,h)_{\phi,p}.$$

The proof of this lemma is very easily by combining (3), (4) and lemma 2.2.

Lemma 2.4. Let $f \in C_{\phi,p}^{r,\mu}(I), 1 \le p \le \infty, n \in \mathbb{N},\ r,s,k > 0,$ and $\delta > 0$, then

(i) $$\Omega_{k,s}^{\mu}(f,\delta)_{\phi,p} \le M_1 2^{s-r}\Omega_{k,r}^{\mu}(f,\delta)_{\phi,p}, \quad s > r > 0,$$

and

(ii) $$\Omega_{k,r}^{\mu}(f,n\delta)_{\phi,p} \le M_2 n^{k-r}\,\Omega_{k,r}^{\mu}(f,\delta)_{\phi,p},$$

where M_1, M_2 are constants independent of f, δ and n.

Proof. Let $f \in C_{\phi,p}^{r,\mu}(I)$, and

$$\Delta_{r,h}^{(\alpha,\beta)} f(x) = \sum_{i=0}^{\infty}(-1)^i f(x)\binom{\frac{r}{2}}{i}\left(A_h^{(\alpha,\beta)}\right)^i,$$

and by (2)-(4), (10) and from lemma 2.2 then

$$\left\|\Delta_{s,\delta}^{(\alpha,\beta)}(\Delta_{r,h}^{(\alpha,\beta)} f)\right\|_{\phi,p} \le 2^{s-r}\left\|\Delta_{r,\delta}^{(\alpha,\beta)}(\Delta_{r,h}^{(\alpha,\beta)} f)\right\|_{\phi,p}$$
$$\le C\,2^{s-r}\,w_r\left(\Delta_{r,h}^{(\alpha,\beta)} f,h\right)_{\phi,p} \le M_1 2^{s-r}\Omega_{k,r}^{\mu}(f,h)_{\phi,p}.$$

Therefore, (i) is satisfied by taking the first supremum when $0 < h \le \delta$, and second one when $0 < t \le h$. In similar way, we can prove the inequality (ii).

3. Main Results

Based on the concepts and strategies of approximation methods in [3, 9, 10-14] and the results in sections 1, we establish the direct and inverse theorems by estimating the best approximation of functions in the Hölder spaces by applying Fourier-Jacobi operators to the fractional differential operator with moduli of smoothness.

Theorem 3.1. Let $f \in C^{r,\mu}_{\phi,p}(I), 1 \leq p \leq \infty, 0 < \mu < r$ and $k > 0$ for algebraic polynomial $P_n \in \Pi_{n-1}$ then

$$E(f)_{C^{r,\mu}_{\phi,p}(I)} \leq C\Omega^{\mu}_{k,r}(f,t)_{C^{r,\mu}_{\phi,p}(I)},$$

where C is a constant not depend on n and f.

Proof. Let $f \in L_{\phi,p}(I), 1 < p \leq \infty$ and consider the de la Vallée Poussin mean [15] as follow

$$\mathcal{V}^{\eta}_n(f)(x) = \sum_{\eta=0}^{n} n^{-1}\eta^2 \mathcal{F}^{(\alpha,\beta)}_{\eta}(f)\mathcal{N}^{(\alpha,\beta)}_{\eta} J^{(\alpha,\beta)}_{\eta}(x), \tag{12}$$

where $\eta = \eta(x)$ is a smooth function such that

$$\eta(x) = \begin{cases} 1 & , |x| < 1/2, \\ 0 & , |x| \geq 1/2. \end{cases}$$

From (5) and Lemma 2.1, we get

$$\begin{aligned} E(f)_{C^{r,\mu}_{\phi,p}(I)} &\leq \|f - \mathcal{V}^{\eta}_n(f)\|_{C^{r,\mu}_{\phi,p}(I)} = \|f - \mathcal{V}^{\eta}_n(f)\|_{\phi,p} + |f - \mathcal{V}^{\eta}_n(f)|_{C^{r,\mu}_{\phi,p}(I)} \\ &\leq Cw_{k+r}(f, n^{-1})_{\varphi,p} + |f - \mathcal{V}^{\eta}_n(f)|_{C^{r,\mu}_{\phi,p}(I)}. \end{aligned} \tag{13}$$

We will estimate the second term of the right-hand side in (13)

$$\begin{aligned} |f - \mathcal{V}^{\eta}_n(f)|_{C^{r,\mu}_{\phi,p}(I)} &\leq \sup_{h>0} \frac{w_r(f - \mathcal{V}^{\eta}_n(f), h)_{\phi,p}}{h^{\mu}} \\ &= \sup_{0<h<n^{-1}} \frac{w_r(f - \mathcal{V}^{\eta}_n(f), h)_{\phi,p}}{h^{\mu}} + \sup_{h \geq n^{-1}} \frac{w_r(f - \mathcal{V}^{\eta}_n(f), h)_{\phi,p}}{h^{\mu}} = I_1 + I_2. \end{aligned} \tag{14}$$

By (3), Lemma 2.1 and (7), we get

$$\begin{aligned} I_2 &= n^{\mu} w_r(f - \mathcal{V}^{\eta}_n(f), n^{-1})_{\phi,p} = n^{\mu} \left\| \Delta^{(\alpha,\beta)}_{r,n^{-1}}(f - \mathcal{V}^{\eta}_n(f)) \right\|_{\phi,p} \\ &\leq Cn^{\mu} \|f - \mathcal{V}^{\eta}_n(f)\|_{\phi,p} \leq Cn^{\mu} w_{k+r}(f, n^{-1})_{\phi,p} \\ &\leq Cn^{\mu} \sup_{0<\delta\leq n^{-1}} \left\| \Delta^{(\alpha,\beta)}_{k,\delta}(\Delta^{(\alpha,\beta)}_{r,\delta} f) \right\|_{\phi,p} \\ &\leq Cn^{\mu} \sup_{0<h\leq n^{-1}} \sup_{0<\delta\leq n^{-1}} \left\| \Delta^{(\alpha,\beta)}_{k,\delta}(\Delta^{(\alpha,\beta)}_{r,h} f) \right\|_{\varphi,p} \leq C\Omega^{\mu}_{k,r}(f,t)_{C^{r,\mu}_{\varphi,p}(I)}. \end{aligned} \tag{15}$$

$$\begin{aligned} I_1 &= \sup_{0<h<n^{-1}} \frac{w_r(f - \mathcal{V}^{\eta}_n(f), h)_{\phi,p}}{h^{\mu}} \leq \sup_{0<h<n^{-1}} \frac{\left\| \Delta^{(\alpha,\beta)}_{r,n^{-1}}(f - \mathcal{V}^{\eta}_n(f)) \right\|_{\phi,p}}{h^{\mu}} \\ &= \sup_{0<h<n^{-1}} \frac{\left\| \Delta^{(\alpha,\beta)}_{r,n^{-1}} f - \Delta^{(\alpha,\beta)}_{r,n^{-1}}(\mathcal{V}^{\eta}_n(f)) \right\|_{\phi,p}}{h^{\mu}}. \end{aligned}$$

From (2), (4) and (12)

$$\Delta^{(\alpha,\beta)}_{r,h}\left(\mathcal{V}^{\eta}_n(f)\right) = \Sigma^{\infty}_{i=0}(-1)^k \begin{pmatrix} \frac{r}{2} \\ i \end{pmatrix} \left(A^{(\alpha,\beta)}_h\right)^i \mathcal{V}^{\eta}_n(f) = \mathcal{V}^{\eta}_n\left(\Delta^{(\alpha,\beta)}_{r,n^{-1}} f\right).$$

Also, lemma 2.1 implies

$$I_1 \leq \sup_{0<h<n^{-1}} \frac{\left\| \Delta^{(\alpha,\beta)}_{r,n^{-1}} f - \mathcal{V}^{\eta}_n\left(\Delta^{(\alpha,\beta)}_{r,n^{-1}} f\right) \right\|_{\phi,p}}{h^{\mu}} \leq C\Omega^{\mu}_{k,r}(f,t)_{C^{r,\mu}_{\phi,p}(I)}. \tag{16}$$

Thus, by combining $(13) - (16)$, then the theorem is proved.

In the following theorem, we shall prove the inverse results.

Theorem 3.2. Let $\in C^{r,\mu}_{\phi,p}(I), 1 \leq p \leq \infty, 0 < \mu < min(r,k)$, then

$$\Omega^{\mu}_{k,r}(f,t)_{\phi,p} \leq C\, n^{\mu-k} \sum_{\iota=0}^{m} \tau^{\iota(r-\mu)} E_{\tau^{\iota}}(f)_{C^{r,\mu}_{\phi,p}(I)},$$

where $\tau = 2$ and $\mathcal{C}$ is a constant which does not depend on n and f.
Proof. Let $g_n \in \Pi_{n-1} \cap C^{r,\mu}_{\phi,p}(I), n \in \mathbb{N}$ be the polynomial of approximation function $f \in C^{r,\mu}_{\phi,p}(I)$ for any $N \in \mathbb{N} \cup \{0\}$ such that $\tau^{N-1} \leq n < \tau^N$.
Suppose that $\sum_{\iota=N}^{\infty} \iota^{n-1} E_\iota(f)_{\phi,p}$ is finite, then we consider

$$f \cong g_{\tau^N} + \sum_{\iota=N}^{\infty}(g_{\tau^{\iota+1}} - g_{\tau^\iota}). \tag{17}$$

From (17) and Lemma 2.3 we get

$$\Omega^\mu_{k,r}(f, n^{-1})_{\phi,p} \leq \Omega^\mu_{k,r}\left(f - g_{\tau^{N+1}}, n^{-1}\right)_{\phi,p} + \Omega^\mu_{k,r}\left(g_{\tau^{N+1}}, n^{-1}\right)_{\phi,p}. \tag{18}$$

Then by applying the Hölder inequality and (6), we get

$$E_n(f)_{C^{r,\mu}_{\phi,p}(I)} = \inf_{P_n \in \Pi_{n-1}} \|f - P_n\|_{C^{r,\mu}_{\phi,p}(I)} = \inf_{P_n \in \Pi_{n-1}} \left\{\|f - P_n\|_{\phi,p} + |f - P_n|_{C^{r,\mu}_{\phi,p}(I)}\right\}$$
$$\geq \inf_{P_n \in \Pi_{n-1}} \left\{ \|f - P_n\|_{\phi,p} \right\} + \inf_{P_n \in \Pi_{n-1}} \left\{ |f - P_n|_{C^{r,\mu}_{\phi,p}(I)} \right\}.$$

Choose $n = \tau^{N+1}, P_n = g_{\tau^{N+1}}$, and from (5), (8) then we get

$$E_{\tau^{N+1}}(f)_{C^{r,\mu}_{\phi,p}(I)} \geq \inf_{g_{\tau^{N+1}} \in \Pi_{n-1}} \left\{ |f - g_{\tau^{N+1}}|_{C^{r,\mu}_{\phi,p}(I)} \right\} = |f - g_{\tau^{N+1}}|_{C^{r,\mu}_{\phi,p}(I)}$$
$$\geq \sup_{0<h<n^{-1}} \frac{w_k\left(\Delta^{(\alpha,\beta)}_{r,n^{-1}}(f - g_{\tau^{N+1}}), h\right)_{\phi,p}}{h^\mu} + \sup_{h \geq n^{-1}} \frac{w_k\left(\Delta^{(\alpha,\beta)}_{r,n^{-1}}(f - g_{\tau^{N+1}}), h\right)_{\phi,p}}{h^\mu}$$
$$= \Omega^\mu_{k,r}\left(f - g_{\tau^{N+1}}, n^{-1}\right)_{\phi,p} + \sup_{h \geq 1/n} \frac{w_r\left(\Delta^{(\alpha,\beta)}_{r,n^{-1}}(f - g_{\tau^{N+1}}), h\right)_{\phi,p}}{h^\mu}.$$

Then we have

$$\Omega^\mu_{k,r}\left(f - g_{\tau^{N+1}}, n^{-1}\right)_{\phi,p} \leq E_{\tau^{N+1}}(f)_{C^{r,\mu}_{\phi,p}(I)}. \tag{19}$$

By theorem 3.1, (6), (10) and (11) we obtain

$$\Omega^\mu_{k,r}\left(g_{\tau^{N+1}}, n^{-1}\right)_{C^{r,\mu}_{\phi,p}(I)} = \sup_{0<h<n^{-1}} \frac{w_k\left(\Delta^{(\alpha,\beta)}_{r,h} g_{\tau^{N+1}} \ , h\right)_{\phi,p}}{h^\mu},$$
$$\leq \mathcal{C} \sup_{0<h\leq n^{-1}} \frac{\mathcal{K}_r\left(\Delta^{(\alpha,\beta)}_{r,h} g_{\tau^{N+1}}, h\right)_{\phi,p}}{h^\mu} \leq \mathcal{C} n^{\mu-r} \left\|\mathcal{D}^r g_{\tau^{N+1}}\right\|_{\phi,p}$$
$$\leq \mathcal{C} n^{\mu-r} \left(\|\mathcal{D}^r g_2 - \mathcal{D}^r g_1\|_{\phi,p} + \sum_{\iota=1}^{N} \|\mathcal{D}^r g_{\tau^{\iota+1}} - \mathcal{D}^r g_{\tau^\iota}\|_{\phi,p}\right)$$
$$\leq \mathcal{C} n^{\mu-r} \left(\|g_2 - g_1\|_{C^{r,\mu}_{\phi,p}(I)} + \sum_{\iota=1}^{N} \tau^{\iota(r-\mu)} \left\|g_{\tau^{N+1}} - g_{\tau^\iota}\right\|_{C^{r,\mu}_{\phi,p}(I)}\right)$$
$$\leq \mathcal{C} n^{\mu-r} \left(E_1(f)_{C^{r,\mu}_{\phi,p}(I)} + \sum_{\iota=1}^{N} \tau^{\iota(r-\mu)} E_{\tau^\iota}(f)_{C^{r,\mu}_{\phi,p}(I)}\right). \tag{20}$$

Choose N such that

$$\tau^N \leq n < \tau^{N+1}.$$

Then from (18), (19), and (20), the required estimation of this theorem follows.
In the following theorem, we present the case which has the sharp results with specific condition in $C^{r,\mu}_{\phi,p}$ −space.

Theorem 3.3. Let $f \in C^{r,\mu}_{\phi,p}(I), 1 \leq p \leq \infty, 0 < \mu < r, 0 < M < \infty$, and $i \geq \mu$. If the inequality $\Omega^\mu_{i,r}(f, h)_{\phi,p} \leq M\Omega^\mu_{j,r}(f, h)_{\phi,p}$ then the following assertion

$$\Omega^\mu_{k,r}(f, n^{-1})_{\phi,p} \leq \mathcal{C} E_n(f)_{C^{r,\mu}_{\phi,p}(I)}, \tag{21}$$

is satisfied for $h > 0$, $j > i$ and a positive constant $\mathcal{C}$.

Proof. Let the condition $\Omega^\mu_{i,r}(f, h)_{\phi,p} \leq M\Omega^\mu_{j,r}(f, h)_{\phi,p}$ be satisfied. By using the inequalities in lemma 2.4 (see [7,14]), we get

$$\Omega^\mu_{k,r}(f, n\delta)_{\phi,p} \leq M\, \Omega^\mu_{k,r}(f, \delta)_{\phi,p}, n \in \mathbb{N}, \delta > 0. \tag{22}$$

IOP Conf. Series: Materials Science and Engineering 1095 (2021) 012010 doi:10.1088/1757-899X/1095/1/012010

And then,

$$\Omega^{\mu}_{k,r}(f,\lambda\delta)_{\phi,p} \le M\Omega^{\mu}_{k,r}(f,\delta)_{\phi,p},$$

we want to prove

$$n^{\mu-\ell}\sum_{\iota=1}^{n}\iota^{\ell-\mu-1}E_{\iota}(f)_{C^{r,\mu}_{\phi,p}(I)} \le M\Omega^{\mu}_{k,r}(f,n^{-1})_{\phi,p}. \tag{23}$$

In fact, we can use theorem 3.1 and (22), to obtain

$$n^{\mu-\ell}\sum_{\iota=1}^{n}\iota^{\ell-\mu-1}E_{\iota}(f)_{C^{r,\mu}_{\phi,p}(I)} \le n^{\mu-\ell}\sum_{\iota=1}^{n}\iota^{\ell-\mu-1}\Omega^{\mu}_{k,r}(f,\iota^{-1})_{\phi,p}$$

$$\le \mathcal{C}Mn^{\mu-\ell}\Omega^{\mu}_{k,r}(f,n^{-1})_{\phi,p}\sum_{\iota=1}^{n}\iota^{\ell-\mu-1} \le \mathcal{C}M\Omega^{\mu}_{k,r}(f,n^{-1})_{\phi,p}.$$

Subsequently, by using theorem 3.2 and the equation (23), for all $m,n \in \mathbb{N}$, $\sigma = mn$, we get the following

$$\Omega^{\mu}_{k,\ell}(f,\sigma^{-1})_{\phi,p} \le \mathcal{C}\sigma^{\mu-\ell}\sum_{\iota=1}^{\sigma}\iota^{\ell-\mu-1}E_{\iota}(f)_{C^{r,\mu}_{\phi,p}(I)}$$

$$= \mathcal{C}\sigma^{\mu-\ell}\left(\sum_{\iota=1}^{n}\iota^{\ell-\mu-1}E_{\iota}(f)_{C^{r,\mu}_{\phi,p}(I)}\ \sum_{\iota=n+1}^{\sigma}\iota^{\ell-\mu-1}E_{\iota}(f)_{C^{r,\mu}_{\phi,p}(I)}\right)$$

$$\le \mathcal{C}\left(Mm^{\mu-\ell}\Omega^{\mu}_{k,\ell}(f,n^{-1})_{\phi,p} + \sigma^{\mu-\ell}\sum_{\iota=n+1}^{\sigma}\iota^{\ell-\mu-1}E_{\iota}(f)_{C^{r,\mu}_{\phi,p}(I)}\right). \tag{24}$$

Inequality (24) implies that

$$\sum_{\iota=n+1}^{\sigma}\iota^{\ell-\mu-1}E_{\iota}(f)_{C^{r,\mu}_{\phi,p}(I)} \ge \frac{\sigma^{\ell-\mu}}{\mathcal{C}}\Omega^{\mu}_{k,\ell}(f,\sigma^{-1})_{\phi,p}.$$

$$-Mm^{\mu-\ell}\Omega^{\mu}_{k,\ell}(f,n^{-1})_{\phi,p}.$$

Now, since $E_n(f)_{C^{r,\mu}_{\phi,p}(I)}$ is a monotonic and the equation (22) is satisfied, then the last equation can be derived as

$$E_n(f)_{C^{r,\mu}_{\phi,p}(I)}\sum_{\iota=n+1}^{\sigma}\iota^{\ell-\mu-1} \ge \left(\mathcal{C}m^{\ell-\mu} - M\right)n^{\ell-\mu}\Omega^{\mu}_{k,\ell}(f,n^{-1})_{\phi,p}.$$

Hence, we can find a positive constant $\mathcal{C} = \mathcal{C}(\ell,\iota,\mathrm{M})$ such that

$$E_n(f)_{C^{r,\mu}_{\phi,p}(I)} \ge \mathcal{C}\Omega^{\mu}_{k,\ell}(f,n^{-1})_{\phi,p}. \tag{25}$$

Therefore, from (25) we obtain (21).

4. Conclusions

We have developed the estimate of the best approximation of the Hölder functions with respect to Fourier-Jacobi operators. In order to get a strong convergent, we have used a special form of the moduli of smoothness to apply for Jacobi harmonic analysis and deal with approximation methods in periodic Hölder spaces. At first, we have introduced two theorems to estimate the best approximation of functions in the Hölder spaces by applying Fourier-Jacobi operators for the direct and inverse cases. Then, we have proved that Jackson-type estimated by moduli of smoothness defined on Hölder spaces. In future study, estimating the approximation of functions in weighted Sobolev spaces by Fourier-Jacobi operators will be considered.

IOP Conf. Series: Materials Science and Engineering 1095 (2021) 012010 doi:10.1088/1757-899X/1095/1/012010

References

[1] Abdulla U 2000 Reaction-Diffusion in Irregular Domains *Journal of Differential Equations* **164** 321-354

[2] Abdulla U & Aal-Rkhais H 2019 Development of the Interfaces for the Nonlinear Reaction Diffusion equation with Convection IOP Conference Series: *Materials Science and Engineering.* 571. 012012. 10.1088/1757-899X/571/1/012012

[3] Abramowitz M & Stegun I A 1965 Handbook of mathematical functions: with formulas, graphs, and mathematical tables (Vol. **55**) Courier Corporation

[4] Bloom W R & Elliott D 1981 The modulus of continuity of the remainder in the approximation of Lipschitz functions *Journal of Approximation Theory* **31** (1), 59-66

[5] Caruntu D I 2007 Classical Jacobi polynomials, closed-form solutions for transverse vibrations *Journal of Sound and Vibration* **306** (3-5) 467-94

[6] Charalambides M & Waleffe F 2006 Spectrum of the Jacobi Tau Approximation for the second Derivative Operator *SIAM J. Numerical Analysis* **46** 280-94

[7] Ditzian Z 2007 Polynomial approximation and $\omega_r^{\phi}(.,t)$ twenty years later *Surveys in Approximation Theory* **3** 106-51

[8] Ditzian Z and Totik T 2012 Moduli of smoothness Vol (**9**) Springer Science & Business Media

[9] Elliott D 1994 On the Hölder semi-norm of the remainder in polynomial approximation *Bulletin of the Australian Mathematical Society* **49** (3) 421-26

[10] Gasper G 1971 Positivity and the convolution structure for Jacobi series *Annals of Mathematics* 112-18

[11] Gautschi W 2004 Orthogonal polynomials Oxford: Oxford university press

[12] Gottlieb D & Shu C W 1997 On the Gibbs phenomenon and its resolution *SIAM review* **39** (4) 644-68

[13] Ioakimidis N I 1983 An improvement of Kalandiya's theorem *Journal of approximation Theory* **38** (4) 354-56

[14] Kadhim H A 2012 On Best Trigonometric Approximation in Lp (μ)-Space *Journal of College of Education for Pure Science* **2** (3) 3-8

[15] Mastroianni G and W Themistoclakis 2008 De la Vallée Poussin means and Jackson's theorem *Acta Scientiarum Mathematicarum* **74** (1-2), 147-70

[16] Platonov S S 2014 Fourier-Jacobi harmonic analysis and approximation of functions *Izvestiya: Mathematics* **78** (1) 106

[17] Rustamov K P 1995 Moduli of smoothness of higher orders related to the Fourier-Jacobi expansion and the approximation of functions by algebraic polynomials *Doklady Mathematics-Interperiodica Translation* **52** (2) 244-247

[18] Szeg G 1939 Orthogonal polynomials Vol. (**23**) American Mathematical Soc.

[19] Tchiotsop D Wolf D Louis-Dorr V & Husson R 2007 ECG data compression using Jacobi polynomials *29th Annual International Conference of the IEEE Engineering in Medicine and Biology Society* (pp. 1863-67)

[20] Weinmann A & Lasser R 2011 Lipschitz spaces with respect to Jacobi translation *Mathematische Nachrichten* **284**(17-18) 2312-26

Study the Structure and Optical Properties of GNPs Doped WO_3 /PS by Spray Pyrolysis Deposition (SPD)

Saadallah F. Hasan [1], Abdul-Majeed E Al-Samarai[2] A. S. Obaid[3], Asmiet Ramizy [4]
[1]Ministry of Education, Direction of Education in AL-Anbar, Anbar, Iraq.
[2]Department of Physics, College of Education for Pure Sciences, University of Tikrit, Tikrit, Iraq.
[3,4]Department of Physics, College of Science, University of Anbar, Anbar, Iraq.

Corresponding author: Saadfy69 @yahoo.com, majeedsa2004@gmail.com,

ahmed.s.obaid.alqayssei@gmail.com, asmat_hadithi@uoanbar.edu.iq

Abstract. In this studied gold (Au) nanoparticles doped tungsten oxide (WO_3) thin film was deposited on porous silicon (PS) substrate at the substrate temperature of 250°C by spray pyrolysis deposition method, with different Au doping concentration of (0, 1, 3, and 5%). The gold nanoparticles GNPs were prepared by using normal atmospheric air cold plasma. All films were annealed at 500°C for 1h. The X-ray diffraction peaks indicated that all films are polycrystalline with a hexagonal structure. The surface morphology was studied by AFM. The average grain size for pure thin film WO_3 was about 61 nm, and it decreased to be about 43.2 nm with increases the doping ratios for the film deposited at 5% of Au. FESEM image all doped and undoped films showed homogenous pattern structure on the porous silicon. The thickness of films was obtained from the cross-section of the FESEM, which was 323±5 nm. A blue-shifted showed based on the photolumenses (PL) peak position calculated of WO_3 as doping ratio increased, the bandgap for Au: WO_3 films lies between (2.85 to 3.02) eV.

Keywords: Thin films, Tungsten Oxide, Gold, Chemical spray pyrolysis, Porous silicon.

1. Introduction

In the past few years, the interest in tungsten trioxide (WO_3) increased quickly and remarkably potential materials and applications in Photovoltaic, solar cell and gas detection, etc. [1]. Among the electrochromic materials, WO_3 is more interesting because of its good lifetime, high-quality look, low constant time, affordable price, and nontoxic property [2,3]. The physical properties of a substance are greatly influenced by its structural arrangement and morphology. [4]. WO_3 is regarded as an n-type semiconductor with stable physicochemical properties [5]. It had a transparency of more than 80% in the visible region, and a wide bandgap (Eg) ranged from (2.5–3.2) eV. [6]. Among noble metals, Au nanoparticles possess many characteristics, because of, it has modern magnetic, chemical, and physical properties, WO_3 films can be elaboration via adding tiny amounts of GNPs into the pure WO_3 [7]. Gold nanostructures are frequently applied because of their size-dependent effects such as high optical absorbance due to surface plasmon resonance [8, 9]. Moreover, GNPs as metal nanostructures are conductive materials that have a significant effect on charge transferring. Therefore, GNPs with excellent stabilities can modify the electrochromic and electrochemical properties of WO_3 thin films. Porous silicon (PS) is a material with a nanostructure that is prepared through electrochemical methods by crystalline silicon. in the mid-1950s, Uhlir discovered porous silicon accidentally at Bell Laboratories [10]. PS possess many characterizations such as the ratio of large surface area to volume, simple fabrication, active surface chemistry, Also it is easily used as the microelectronic, Micro-

Electro-Mechanical Systems (MEMS), and applications like batteries, sensors electronics, optics, biomedicine, and solar cells. [11-13]. WO_3 can be prepared in several ways. The most important ways are the thermal evaporation [14], electrodeposition method [15], a sputtering technique [16], sol-gel [17], CVD [18], and spray pyrolysis deposition [19]. Attention has been focused on the SPD method, as it can simply prepare the films in the air. Therefore, it is considered one of the most attractive methods for preparing thin films due to a large surface area coating that can be obtained by used low-cost equipment with minimum energy consumption, and it allows the utilization of inexpensive precursor materials [22-20]. Essentially, SPD depends on a pyrolysis reaction on a heated substrate of the precursor, which spread as small droplets. Because of adding the surfactant of the precursor, the surface tension will be minimize, which let the figuration of small droplets above the substrate. Else, is the temperature of droplet increase more when approaching of the substrate that continuously adjusts the reaction rate [23]. The aim of the study is to study the effect of Au GNPs doping on the structural and optical properties of WO_3 thin film prepared by SPD.

2. Experimental

2.1 Preparation of WO_3 solution by Chemical Reaction

To obtain tungsten oxide, 5g of tungsten powder (W) with purity of (99.9%) was dissolved in 100 ml of hydrogen peroxide at a concentration of 30%. The reaction was carried out in a water bath at a temperature of 25°C until the reaction was complete for 10 min, and then the excess of hydrogen peroxide was eliminated at a temperature 150°C for 15 min in a magnetic mixer. Then 20ml of ethanol was added to reaction solution at a temperature of 70°C for a period of (1h) in a magnetic stirred with cover, but not tightly the beaker because the reaction is accompanied with strong heat.

2.2 Preparation of WO_3 thin film

Thin WO_3 can be deposited via spraying the solution on a hot substrate (PS) at a temperature of about 250°C. After the deposition process, the tungsten film left on a hot base for at least one hour to form the oxidation process and complete the crystal growth. The thermal spray method was used to prepare the films with a concentration of 0.21 M solution and a pressure of 1 bar, and the distance between the base and the nozzle was about 32 cm, the deposition rate was 10:15 sec

2.3 Preparation of GNPs solution

The salts of aqueous tetrachloride $HAuC_{l4.4}H_2O$ which is used as a range of purity (99%) are produced by SGMA (German company). The prepared quantity is 20 ml volume while the required weight is computing by the following equation:

Concentration (mole) = (mass (gram)) / (Molecular weight (g / mol) × volume (litter))

The gold salts solution was prepared in the required volume and a concentration of 1 mM, there after the argon gas tube is opened, the metal tube with a diameter of 1 mm is fixed vertically by the catcher, then the prepared solution was placed under the metal tube. The distance between the tube nozzle and the solution surface becomes 1mm when the beaker getting close to the metal tube. The inter gas quantity inside the metal tube was controlled by flow meter of speedometer and the gas tube. The voltage produced by the system gradually increases till the case of the plasma generated between the surface of the fluid and the tube.

3. Results and Discussion

3.1. X-ray diffraction

XRD pattern of pure WO_3 film and doped deposited WO3 film with Au NPs on PS at substrate temperature of 250°C illustrated in fig.1. From the figure, it is observed the many diffraction peaks which indicated that the structure is polycrystalline for the pure and doped films, the planes (001), (110), (111), (300) and (002), at 2θ =22.79^0, 24.388^0, 28.42^0, 33.68^0, 42.95^0, 46.59^0 respectively, were obsorved the preferred peak at (110) plane. The diffraction peaks and their relative intensity exactly match that of JCPDS 00-033-1387 for WO_3 hexagonal. This result corresponding with Ibrahim et al [24]. The peak at 2θ=28.42^0 Indicate to silicon at (111) plane. When doped of Au at different concentrations, it was found that the intensity of the peaks of WO_3 began to gradually decrease towards fewer angles with increase doping ratio. This result may be due to the substituted of Au by lattice sites within WO_3. The doping with Au did not transform the orientation of WO_3 thin film crystalline outgrowth for the dominant planes, and outgrowth continues through the direction of (110), according to the drift competitive growth model, which is called "the model of the survival of the fastest-growing crystallites [25]. By Debye-Scherrer equation, the average crystallite size of the crystal is calculated based on the preferred peaks present in the XRD pattern [26],

$D = k \lambda / \beta \cos\theta$ (1)

Where λ is the wavelength, β is a full-width half of the maximum, θ is the Bragg angle, and (k) is form factor equal to 0.9. The result showed that the crystalline size decrease from (23.37 to 14.27) nm with increase doping ratios.

3.2. Atomic force microscope (AFM)

3-D WO_3:Au/PS AFM images as shown in Fig. 2, display the topography of the WO_3 surface, which agreement with XRD result. Average diameter, roughness, and R.M.S of samples are illustrated in table 1. The pure WO_3 specimen has a granular structure, fully ordered, and closely packed on the surface. After doping it was found that the grain size, roughness and RMS decrease and this is agreement with [24] as showed in table 1.

3.3. Field Emission Scanning Electron Microscopy (FESEM) and EDX

Figure. 3 (a, b, c and d) showed the FESEM images of WO_3/PS doped with Au at 0,1,3,5%vol., respectively. It was clearly observed that all the samples showed approximately spherical-shaped particles, and having granulation and a tendency to agglomerate with uniform distributed size morphology with an average diameter of (22–97) nm. The images in agreement with another researcher group [27]. The crystalline size evaluated from XRD results so smaller than the FESEM, which decreases with increasing Au ratio. Fig (3-e) showed the thickness from the cross-section of PS and WO_3 thin film, which was about 21μm and 323±5 nm respectively, it displays the formation of nonparallel and partially cracked silicon walls, which result in asymmetrical and tight holes between them [28].

EDX analysis was showed in figure 4, which indicated the EDX analysis of the WO_3 nanoparticles that were produced by the spray pyrolysis method. The results exhibit the characteristic peaks of W and O present in the film. The peak intensity of W was reduced may be of the addition of Au ions at the site of W, and the O intensity peak stays constant for both pure and Au doped WO_3.

3.4. Photoluminescence

PL spectroscopy is one of the most important studies to mensuration the energy distribution of the photons emitted after optical excitation and to comprehend the electron-hole pair in semiconductor oxide. Using photoluminescence spectra, oxygen vacancy, and lattice deformation were analyzed for samples of pure tungsten oxide and Au NPs doped were illustrated in figure 5. Both pure WO_3 and doped thin film with Au exhibit blue emission as compared with the bulk WO_3. The energy gap increases as function of Au –doped this blue-shifted is due to a reduction to nano size, which the supports charge carrier quantum confinement model [29]. The low intensity of PS and the thin film when doped, may be because the gold created traps to capture non-radioactive electrons, which reduced the intensity of emission and the loss of electron energy, which recombination into heat and not light. Conversely, there is a blue-shift peak towards higher energy, indicating a decrease in crystal size and thus an increase in the energy gap.

4. Conclusion

This paper demonstrates using spry method as easy and inexpensive method to synthesize pure WO_3 and doped with Au. XRD analyses that the crystal size decreases due to an increased doping ratio. AFM topography confirmed that the grain size in nanoscale, and it decreases with increase doping ratios. From FESEM, it was observed that all the samples showed approximately spherical-shaped particles with a porous structure having granulation and a tendency to agglomerate with a uniform distributed size. The energy dispersive spectrum (EDX) underlines the presence of W and O in the prepared materials. PL results showed a blue shift with an increased Au doping ratio in the prepared thin films.

References

[1] Mitsugi, F., Hiraiwa, E., Ikegami, T., Ebihara, K., & Thareja, R. K. (2002). WO3 thin films prepared by pulsed laser deposition. Japanese journal of applied physics, 41(8R), 5372. P. E. Hodgson, "Nuclear Reactions and Nuclear Structure", Clarendon Press, Oxford, (1971);

[2] Pecquenard, B., Lecacheux, H., Castro-Garcia, S., & Livage, J. (1998). Electrochromic properties of peroxopolytungstic acid thin films. Journal of sol-gel science and technology, 13(1-3), 923-927. C. Kalbach, Phys. Rev. C47 (1993) 587;

[3] Heckner, K. H., & Kraft, A. (2002). Similarities between electrochromic windows and thin film batteries. Solid State Ionics, 152, 899-905. C. Kalbach, "User's Manual for PRECO-2006, Exciton Model Preequilibrium Nuclear Reaction Code with Direct Reactions", Triangle Universities Nuclear Laboratory, Duke University, (2007);

[4] Nishioka, D., Tsuchiya, T., Higuchi, T., & Terabe, K. (2020). Oxygen-tolerant operation of all-solid-state ionic-gating devices: advantage of all-solid-state structure for ionic-gating. Japanese Journal of Applied Physics, 59(SI), SIIG09. J. Csikai&S.Nagy, Acta Physica 21 (1966) 303;

[5] GUO, X. Z., KANG, Y. F., YANG, T. L., & WANG, S. R. (2012). Low-temperature NO2 sensors based on polythiophene/ WO3 organic-inorganic hybrids. Transactions of Nonferrous Metals Society of China, 22(2), 380-385. W. Scobel, M. Blann, T. T. Komoto, M. Trabandt, S. M. Grimes, L. F. Hansen, C. Wong, B. A. Pohl, Phys. Rev. C30 (1984) 1480;

[6] Simchi, H., McCandless, B. E., Meng, T., & Shafarman, W. N. (2014). Structural, optical, and surface properties of WO_3 thin films for solar cells. *Journal of alloys and compounds*, *617*, 609-615.P. Guazzoni, L. Zetta, P. Demetriou, P. E. Hodgson, Z. Phys. A — Hadrons and Nuclei 354 (1996) 53;

[7] Ibrahim, I. M. (2018). The effect of gold nanoparticles on WO_3 thin film. *Iraqi Journal of Physics (IJP)*, *16*(36), 11-28.I. C. Sagrado Garcia, J. F. Lecolley, F. R. Lecolley, V. Blideanu, G. Ban, J. M. Fontbonne, G. Itis, J. L. Lecouey, T. Lefort, N. Marie, J. C. Steckmeyer, C. Le Brun, J. Blomgren, C. Johansson, J. Klug, A. Orhn, P. Mermod, N. Olsson, S. Pomp, M. Osterlund, U. Tippawan, A. V. Prokofiev, P. Nadel-Turonski, M. Fallot, Y. Foucher, A. Guertin, F. Haddad, M. Vatre, Phys. Rev. C 84 (2011) 044619;.

[8] He, T., Ma, Y., Cao, Y., Yang, W., & Yao, J. (2001). Enhanced electrochromism of WO_3 thin film by gold nanoparticles. Journal of Electroanalytical Chemistry, 514(1-2), 129-132.

[9] Nagai, J., McMeeking, G. D., & Saitoh, Y. (1999). Durability of electrochromic glazing. Solar energy materials and solar cells, 56(3-4), 309-319.

[10] Uhlir Jr, A. (1956). Electrolytic shaping of germanium and silicon. Bell System Technical Journal, 35(2), 333-347.

[11] Levitsky, I. A. (2015). Porous silicon structures as optical gas sensors. Sensors, 15(8), 19968-19991.

[12] Pérez, M., Dutt, A., de la Mora, B., Mon-Pérez, E., Villagrán-Muniz, M., García-Sánchez, M. F., & Santana, G. (2018). Effect of ammonia plasma treatment on the luminescence and stability of porous silicon. Materials Letters, 216, 277-280.

[13] Lukianov, A., & Ihara, M. (2018). Free-standing epitaxial silicon thin films for solar cells grown on double porous layers of silicon and electrochemically oxidized porous silicon dioxide. Thin Solid Films, 648,1-7.

[14] Lozzi, L., Ottaviano, L., Passacantando, M., Santucci, S., & Cantalini, C. (2001). The influence of air and vacuum thermal treatments on the NO2 gas sensitivity of WO_3 thin films prepared by thermal evaporation. Thin Solid Films, 391(2), 224-228.

[15] Vijayalakshmi, R., Jayachandran, M., & Sanjeeviraja, C. (2003). Structural, electrochromic and FT-IR studies on electrodeposited tungsten trioxide films. Current Applied Physics, 3(2-3), 171-175.

[16] Purans, J., Kuzmin, A., Parent, P., & Laffon, C. (2001). X-ray absorption study of the electronic structure of tungsten and molybdenum oxides on the O K-edge. Electrochimica acta, 46(13-14), 1973-1976.

[17] Cantalini, C., Wlodarski, W., Li, Y., Passacantando, M., Santucci, S., Comini, E., ... & Sberveglieri, G. (2000). Investigation on the O3 sensitivity properties of WO_3 thin films prepared by sol–gel, thermal evaporation and rf sputtering techniques. Sensors and Actuators B: Chemical, 64(1-3), 182-188.

[18] Sivakumar, R., Raj, A. M. E., Subramanian, B., Jayachandran, M., Trivedi, D. C., & Sanjeeviraja, C. (2004). Preparation and characterization of spray deposited n-type WO3 thin films for electrochromic devices. Materials Research Bulletin, 39(10), 1479-1489.

[19] Yang, T., Zhang, Y., & Li, C. (2014). Large scale production of spherical WO_3 powder with ultrasonic spray pyrolysis assisted by sol–gel method for hydrogen detection. Ceramics International, 40(1), 1765-1769.

[20] Bertus, L. M., Enesca, A., & Duta, A. (2012). Influence of spray pyrolysis deposition parameters on the optoelectronic properties of WO_3 thin films. Thin Solid Films, 520(13), 4282-4290.

[21] J. M. Rzaij, N. F. Habubi, "Room temperature gas sensor based on La2O3 doped CuO thin films", Applied Physics A (2020) 126:560

[22] Bertus, L. M., & Duta, A. (2012). Synthesis of WO_3 thin films by surfactant mediated spray pyrolysis. *Ceramics International*, *38*(4), 2873-2882.

[23] Bertus, L. M., Faure, C., Danine, A., Labrugère, C., Campet, G., Rougier, A., & Duta, A. (2013). Synthesis and characterization of WO3 thin films by surfactant assisted spray pyrolysis for electrochromic applications. Materials chemistry and physics, 140(1), 49-59.

[24] Isam M. Ibrahim1, Niran F. Abdul-Jabbar2, Abeer H. Fezaa, (2018)."The effect of gold nanoparticles on WO3 thin film", Iraqi Journal of Physics Vol.16, No.36, PP. 11-28.

[25] Ramizy, A., Ibrahim, A. S. M. I. M., & Eisa, M. H. (2014). Performance of Multi-Function Devices Fabricated from La2O3-Doped NiO Thin Films.

[26] Mohammed, A. S., Kafi, D. K., Ramizy, A., Abdulhadi, O. O., & Hasan, S. F. (2019). Nanocrystalline Ce-doped CdO thin films synthesis by spray pyrolysis method for solar cells applications. *Journal of Ovonic Research Vol*, *15*(1), 37-42.

[27] Alaei, M., Mahjoub, A. R., & Rashidi, A. (2012). Preparation of different WO_3 nanostructures and comparison of their ability for Congo red photo degradation.

[28] Ibrahim, I. M., Mohammed, A. S., & Ramizy, A. (2018). Responsivity Enhancement of Lutetium Oxide Doped–NiO Thin Films. *Journal of Ovonic Research Vol*, *14*(1), 17-25.

[29] Mahmood, H. M., Abdullah, M. A., Ramizy, A., & Mohammed, A. S. (2020). The effect of spin coating parameters on GaN nanostructures thin films properties deposit on different substrates. *Digest journal of nanomaterials and biostructures*, *15*(2), 587-591.

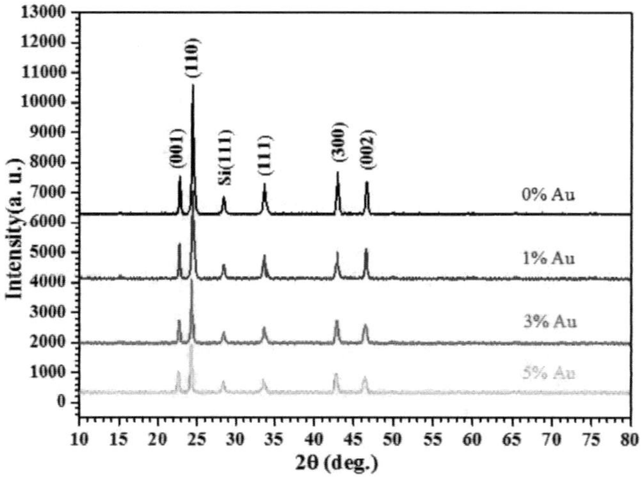

Figure 1: XRD of the WO_3:Au/PS at annealing of (500 °C).

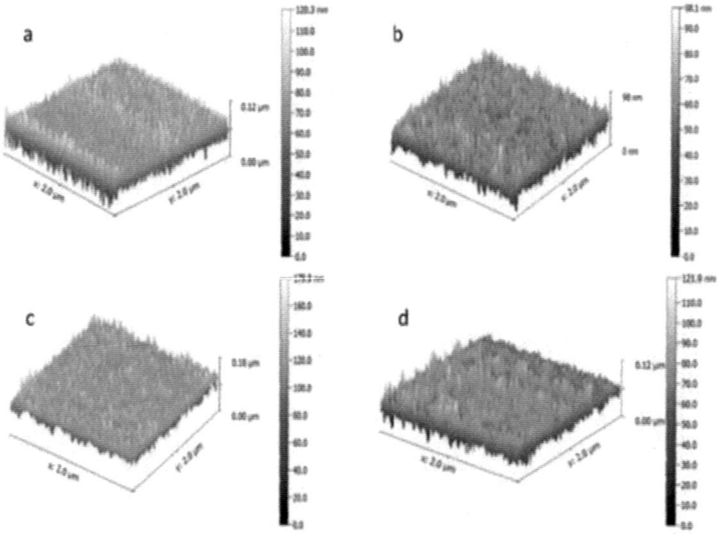

Figure 2: AFM image of WO_3 doped (0, 1, 3, 5,) % ratios of Au at annealing temperature (500 °C).

Table1: AFM parameters of WO_3 doped (0, 1, 3, 5,) % ratios of Au at annealing temperature (500 °C).

Au (%)	Ave. diameter (nm)	Roughness (nm)	R.M.S (nm)
0%	61.0	8.8	11.1
1%	60.1	7.9	10.7
3%	53.3	5.4	6.8
5%	43.2	3.6	4.7

Figure 3: FESEM images of WO_3/PS (a) 0%, (b) 1%, (c) 3%, (d) 5% of Au, (e) cross section of WO_3/PS.

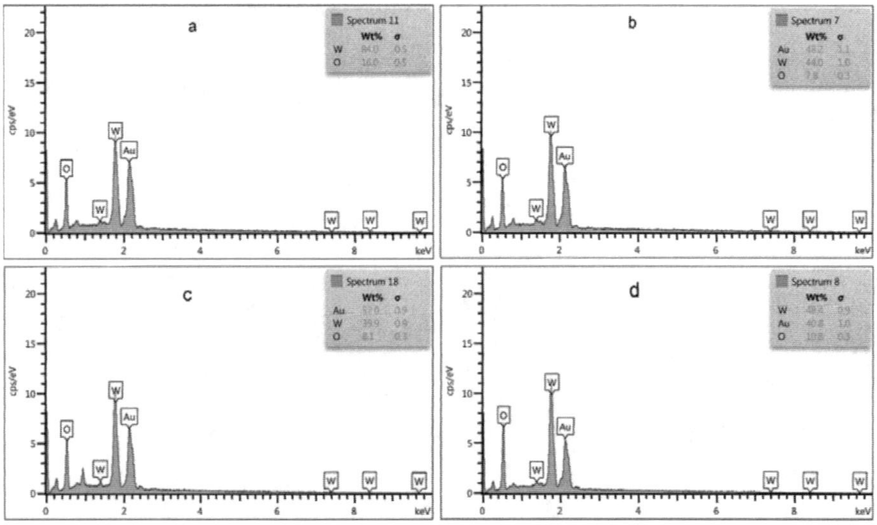

Figure 4: Energy-dispersive X-ray (EDX) spectra of WO3

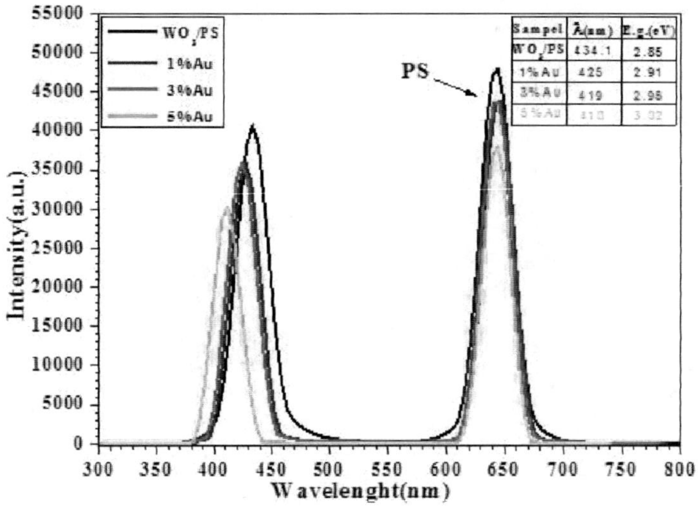

Figure 5: Photoluminescence spectra for WO_3 with different doping concentration

IOP Conf. Series: Materials Science and Engineering 1095 (2021) 012012 doi:10.1088/1757-899X/1095/1/012012

Studying the Nuclear Structure of the ($^{172}_{70}Yb_{102}$) Deformed Nucleus using (IBM-1) and (GVMI) Models

Saad Hameed Al-Fahdawi [1], **Dr.Ali Kalaf Aobaid** [2]

Department of Physics, College of Education for Pure Science, University of Anbar, AL-Anbar, Iraq

Corresponding author: [1] saadhammed19@uoanbar.edu.iq ,[2] esp.alik.obaid@uoanbar.edu.iq

Keywords: Nuclear structure, Intersection, bosons, moment of inertia.

Abstract: In this research, some properties of the nuclear structure of even-even $^{172}_{70}Yb_{102}$ deformed nucleus were studied with dynamic symmetry SU (3).The energy levels E (L), excited energy levels (E_{γ}), and electric quadrupole moment(Q_L) were calculated using a set of parameters in the calculation to obtain a match with the practical values available using the first interacting boson model. IBM-1 and generalized variable moment of inertia (GVMI) model, The calculated results are in good agreement with the available practical values, especially, the values calculated according to the GVMI program. The phenomenon of band intersection ,and back bending were also studied, the results showed no effect of the moment of inertia on the shape of the nucleus under consideration ,because the back bending phenomenon was not occur.

1-Introduction

Several nuclear models study the nuclear structure, they are called "nuclear models", one of them is liquid drop model which proposed by (Von Weizsacker, 1934) [1] which explained the binding energy and the phenomenon of Nuclear fission, but it did not succeed in explaining the stability of the nucleus. As for the shell model proposed by (W. Elsasier, 1935) [2], practical experiments have shown that the stability of the nucleus is high when the number of nucleons is equal to one of the magic numbers [3]. In addition to that, it determined the angular momentum of energy levels, but at the same time failed to explain the nuclear spin of the ground level of the even-even nucleus, which always equals zero [4].

In 1974 Arima ℨ Iachello [5] proposed a nuclear model in which it was able to describe the characteristics of the energy levels in the even-even nuclei and positive parity(π^+) of medium and heavy mass numbers by the pairs of nucleons outside the closed shell that were treated as bosons [6] so that it did not take into account the degrees of freedom for these bosons, they were called the first interacting boson model IBM-1. This model was developed by introducing degrees of freedom to bosons. As a result of this modification, new nuclear properties of the nuclei were revealed and it was called the second interacting boson model IBM-2 [7].

The first interacting boson model does not distinguish between proton bosons (s_π, d_π) and neutron bosons (s_v, d_v). The total number of bosons (N) is calculated from the sum of (N_π proton bosons and N_v neutron bosons) as pairs of particles, Starting from the closest closed shell to the middle of the next shell, meaning that the number of bosons (N) is equal to the number of pairs of particles outside the closed shell, which is a fully preserved and constant quantity for each nucleus, but if it is more than half of the shell then (N_π) and (N_v) is taken as the number of Hole Pairs [8,9].

In this model, we find that the S and d bosons can interact with each other, and as a result, the general formula for the Hamiltonian effect for this system is written after the creation effects (S^+ , d^+) and the annihilation effects (S, d) (Annihilation Operators) are defined. [10, 11]:-

$$H=\varepsilon_s(s^+.\tilde{s})+\varepsilon_d(d^+.\tilde{d})+\sum_{L=0.2,4}\frac{1}{2}(2L+1)^{\frac{1}{2}}C_L[(d^+\times d^+)^{(L)}\times(\tilde{d}\times\tilde{d})^{(L)}]^{(0)}$$
$$+\frac{1}{\sqrt{2}}\tilde{V}_2[(d^+\times d^+)^{(2)}\times(\tilde{d}\times\tilde{s})^{(2)}+(d^+\times s^+)^{(2)}\times(\tilde{d}\times\tilde{d})^{(2)}]^{(0)} \quad \text{---- (1)}$$
$$+\frac{1}{2}\tilde{V}_0[(d^+\times d^+)^{(0)}\times(\tilde{s}\times\tilde{s})^{(0)}+(s^+\times s^+)^{(0)}\times(\tilde{d}\times\tilde{d})^{(0)}]^{(0)}$$
$$+U_2[(d^+\times s^+)^{(2)}\times(\tilde{d}\times\tilde{s})^{(2)}]^{(0)}+\frac{1}{2}U_0[(s^+\times s^+)^{(0)}\times(\tilde{s}\times\tilde{s})^{(0)}]^{(0)}$$

Where C_L (L = 0,2,4), V_L (L = 0,2), U_L (L = 0,2) describes the interactions of the bosons with each other, and these coefficients depend on the number of N bosons. As for the parentheses, they represent the angular momentum. Several other formulas are equivalent to the general formula for IBM-1 of Hamilton, where $\varepsilon = \varepsilon_d + \varepsilon_S$ represent the energy of the bosons. To simplify it, the energy of the bosons was considered to be zero, thus: $\varepsilon=\varepsilon_d$

So we can rewrite the equation (1) using the Multiple Expansion Formula [12]: -

$$\hat{H}=\varepsilon\hat{n}_d+a_o\hat{P}.\hat{P}+a_1\hat{L}.\hat{L}+a_2\hat{Q}.\hat{Q}+a_3\hat{T}_3.\hat{T}_3+a_4\hat{T}_4.\hat{T}_4\text{-------(2)}$$

Where $\mathcal{E}_s$ - $\mathcal{E}_d$= $\mathcal{E}$ represents the energy of the bosons, and for ease assumed that the energy of the boson S is equal to zero. As for the coefficients a_0, a_1, a_2, a_3, a_4, they express the strength of the interaction of duality, the angular momentum, the quadrupole, the octopole, and the sixteenth pole between the bosons respectively.

The general formula for the electric quadrupole transition effect can be written with the following relationship[13]:

$$T_\mu^{(E2)}=\alpha_2[d^\dagger s+s^\dagger d]_\mu{}^{(2)}+\beta_2[d^\dagger d]_\mu{}^{(2)}\text{.............(3)}$$

Based on the values of the energy ratios and their comparison with the typical values, it was found that this core belongs to the rotational determination [14]. So we'll say a little bit about rotational dynamic symmetry.

The second determination presented by (Arima and Iachello) is in which the energy of the boson (ε) is much smaller than the reaction energy (V) i.e. (V >>ε), as (V) represents the energy of the quadrupole momentum reaction (Q.Q) between the bosons. Hamilton's function to determine by the equation[15]:

$$\hat{H}^{(II)}=a_1\hat{L}.\hat{L}+a_2\hat{Q}.\hat{Q}\text{..........(4)}$$

From equation (4) we find that the influences ($\hat{\varepsilon}$, $\hat{P}$, $\hat{T}_3$, $\hat{T}_4$) are ineffective, as we notice that The angular momentum dipole interaction (L.L) in addition to the electrode quadrupole interaction (Q.Q) between bosons are dominant in this determination. As for Hamilton's eigenvalues equation SU (3), it is given by the following relationship [14]:

$$E|N,(\lambda,\mu),K,L,M\rangle = \frac{a_2}{2}(\lambda^2+\mu^2+\lambda\mu+3(\lambda+\mu)+(a_1-\frac{3a_2}{8}).L(L+1)-------(5)$$

The eigenvalues are known as the previously defined quantum numbers (M, L, N) over the two quantum numbers(μ, λ). As they represent the states of SU (3) and the quantum number **K** denotes the states that have equal values for and the figure (2 -2) shows a typical spectrum for the determination of SU(3)[12,15].

The reduced electromagnetic transition probability can be calculated from the relation [16]:

$$B(E2, L_i \rightarrow L_f) = \frac{1}{2Li+1}\left|\left\langle L_f \left\| \hat{T}^{(E2)} \right\| L_i \right\rangle\right| ------------------(6)$$

Where: $\left|\left\langle L_f \left\| \hat{T}^{(E2)} \right\| L_i \right\rangle\right|$ the reduced matrix elements for the electric quadrupole transition probability.

2- Electric quadrupole Momentum:

The moment of the electric quadrupole is another important property of the cores, and is defined as the amount of anomaly or deviation from the symmetric spherical distribution of the electrical charge inside the core. Knowing it gives an idea of the shape of the nucleus and calculates its value through the equation:

$$Q_L = <L,M_L = L\,|\sqrt{\frac{16\pi}{5}}\,T_0^{(E2)}\,|\,L,M_L = L> = \sqrt{\frac{16\pi}{5}}\begin{pmatrix} L & 2 & L \\ -L & 0 & L \end{pmatrix} < Lf\left|\left|T^{\wedge(E2)}\right|\right| L_i > \cdots (7)$$

Where it was found in practice that the value of the moment of the electric quadrupole is equal to zero for the spherical nuclei (closed shell), and as the number of bosons increases, the moment value of the electric quadrupole increases, meaning that the amount of distortion increases until the core becomes permanently deformed, and the nuclear deformation is of two types:

1-Prolate deformation :

In this type, the Nucleus rotates around an axis perpendicular to the axis of nuclear symmetry. This type has a positive Value of the of the electric quadrupole moment ($+Q_L$).

2- Oblete deformation :

In this type, the Nucleus rotates in an axis parallel to the symmetric nuclear axis. This type occurs in spherical cores, and has a negative value of the electric quadrupole moment ($-Q_L$).

3- Generalized Variable Moment of Inertia (GVMI) Model

D. Bonatsos and A. Klein [17] proposed a new model for calculating energy levels of nuclei. This model is more comprehensive than the Variable Moment of Inertia (VMI) Model and the Harmonic Vibration Model. The general formula for energy levels in this model contains the basic functions of (VMI) that change with the change of angular momentum L and L(L+1). The energy levels as a function of angular momentum (L) is [18]:

$$E(L) = \frac{1}{2}\left[\frac{L(L+1)}{\vartheta(L)} + C(\vartheta(L) - \vartheta_0)^2\right] \qquad (8)$$

Where :

$\vartheta(L)$ is a moment of inertia as a function of L, and C, ϑo are parameters which are suitable with fitted to the experimental data.

This equation consists of many terms, D. Bonatsos and A. Klein reduced these terms to two terms[17]. They are based on the same basic functions. The theoretical results obtained from this model are fully consistent with the experimental values of energy levels for SU (3), SU (5) for the nuclei more than those found in the other models.

A.M. Kalaf and others (2020) [19] used a variable moment of inertia model (VMI) to study identical bands (IB's) for mass number nuclei $158 \leq A \leq 170$. As results were obtained that matched the practical values.

4- Energy Band intersection:

The intersection of band means that the moment of inertia of a specific energy level for any band such as (β) or(ɤ) intersects or replaces any energy level of any other band such as a g-band . Between any two band have the same spin but differ in energy and moment of inertia[20].

5- Results and discussion:

In this paper, the properties of the even-even ($^{172}_{70}Yb_{102}$) nucleus were studied (such as energy levels, transition energy Eɤ, and the potential for electrical transfers), the energy levels of this core were arranged in the form of bands according to their appearance and the arrangement of the bands was as follows (g, β1, β2, ɤ1) and this arrangement indicates that the nucleus belongs to the symmetry SU (3) or SU (5), where the arrangement of these bands is very important for its use in calculating the behavior of the selected nuclei and the values of the moment of inertia for all the bands, and through the practical energy ratios it was found that the nucleus ($^{172}_{70}Yb_{102}$) belongs to the symmetry SU (3). The energy levels for this nucleus were also calculated using the first interacting boson model IBM-1 and the GVMI program. The energy levels for IBM-1 were calculated through the best interaction coefficients suitable for Equation

(2) shown in Table (1) of Matches theoretical values (pw) with practical values (exp.). While the GVMI values were calculated for any band of parameters (E_k). These measurements were chosen from the smallest value of chi-square (χ^2)[18] :

$$x^2 = \frac{\left(E_{cal}-E_{exp.}\right)^2}{E_{cal}^2} \dots\dots\dots\dots\dots (9)$$

where the parameters corresponding to the value of chi-square were given in Table (2) for all bands. This table shows that the parameter E_k is equal to zero for the ground band and takes the band head energy for the bands (β_1, β_2). As for the gamma band (γ_1-band), this value is variable.

Figure (1) shows the ground band where there is a match between the values of the practical energy levels and the IBM-1 values for the levels L<8. As for the levels L> 8, there is a convergence between the energy levels, this difference in values is attributed to the IBM-1 model does not distinguish between bosons Protons N_π and bosons N_V neutrons. As for the GVMI moment of inertia model, there is a great match between the theoretical value and the practical values for all levels of the ground band to choose the best parameters for this model.

Figure (2) shows the beta (β1) band where theoretical values for IBM-1 programs and L<6 levels are identical to the practical values. For high levels of L>6, the theoretical values are greater than the practical values. As for the GVMI model values, they are completely in agreement with the practical values for all levels.

Figure (3) shows that there is no good match between the theoretical values of (IBM-1) and the experimental values, as the results agree only at momentum L = 6, while there is a convergence of the GVMI values with the experimental values better than the IBM-1 values, which are greater or smaller than the rest of the values.

Figure (4) shows the calculated results for the gamma band (γ1) where the IBM-1 results are slightly higher than the other results at low levels and far apart at L> 6 levels because this model does not distinguish between proton bosons and neutron bosons. As for the results of GVMI, they agree with the experimental results at specific levels and diverge slightly at other levels.

The phenomenon of band intersection illustrated in Fig. (5) is an important topic because it is used to explain the phenomenon of back bending where bands intersect (β1, β2, γ1) at the angular momentum L = 14. The dissociation of a pair of nuclei crosses the bands and forms a back curvature phenomenon that was not seen in this nucleus. Band intersection also occurs due to a low moment of inertia at large angular momentum. The effect of the Curious force will reduce the energy of the nucleons, and the instability of these two excited states leads to the intersection of one beam with another.

Figure (6) shows the shape of the Nucleus and the amount of Deformation; where it was observed that the shape of the Nucleus takes the shape of an Elongated Oval at Momentum (21+, 23+, 25+, 26+, 28+, 210+). While it takes a flat ellipse shape at Momentum (22+, 24+, 27+, 29+) and has the highest deformation at Angular Momentum (21+) and its Value (1.964 eb).

Figure (7) shows the relationship between the moment of inertia and the square of the rotational energy, Where it is noticed from the figure that there is no back bending in the bundles of this core, which indicates the absence of an effect of the moment of inertia on the shape of the nucleus.

References:

[1] C.F. Von Weizsacker, Z. Phys. Vol. 6, P.431, (1935).

[2] W.J. Elsasser, Phys. Vol. 4, P.549, (1933).

[3] W. Meyerholf, "Elements of Nuclear Phys.", Mc Graw-Hill, (1967).

[4] H. A. Enge. " Introduction to nuclear Phys.", (1983).

[5] F. Iachello, and A. Arima, Phys., Lett. B53, P.309, (1974).

[6] A. Arima, T. Otsuka, F. Iachello, and I. Talmi, Phys. , Lett. B66, P.205, (1977).

[7] R.F. Casten, A. Gelberg and P. Von Brentano, Phys. Rev. C36, P.1225, (1987).

[8] S.H.Trier," A study of nuclear Structure of 160-180Hf even-even Isotopes by the IBM-1.", M. Sc. Thesis, Babylon University, (2006).

[9] O.Scholten, F. Iachello, and A. Arima, "Interacting Bosons Model of Collective Nuclear States III. The Transition Form SU(5) to SU(3) .", Ann.phys.115, (1978).

[10] W.Greiner,J.A.Maruhn, "Nuclear Models.",Spring-Verlag Berlin Heidelberg NewYork,(1996).

[11] F. Iachello," An Introduction to the Interacting Boson Model .", Plenum Press,(1980).

[12] R. Casten and D. Warner, Rev. Mod. Phys. 60, 389, (1988).

[13] A. Arima and F. Iachello , Ann. Phys.,123 ,468 ,(1979).

[14] K. Abrahams, K. Allaart, and A. Dieperink, Dieperink nuclear structure, New York and London,67,53, (1981).

[15] A. Arima, and F. Iachello, Ann. Phys . 111 , 201 , (1978).

[16] J . Jabber , "Decay Schemes From the (n,γ) Reaction on 151Eu and 181Ta" , Ph. D Thesis , London University , (1989).

[17] D. Bonatsos and A. Klein; phys. Rev. C,Vol.29,P.1879 (1984).

[18] M. A. J. Mariscotti , Gertude Scharff – Gold Haber , and Brain Buck ; Phys. Rev. vol. 178 , p.1864 (1969).

[19] A. M Khalaf.,Asmaa Abdelsalam ,"Identical bands in doubly even nuclei in framework of variable moment of inertia (VMI) and interacting boson models" vol 997,(2020).

[20] J .p Elliott, , and A. P. white Phys. Lett. B vol.97.p.169. (1980).

[21] Balraj Singh , Nuclear Data Sheets 75, 199, P.214 (1995).

Table (1): The best fitted interaction parameters of equation (2) for the energies in(MeV)units

E	*P^† .P^*	*L^.L^*	*Q^.Q^*	*T^3 .T^3*	*T^4 .T^4*	*CHI*
0.0001	*0.0000*	*0.0088*	*-0.0112*	*0.0000*	*0.0000*	*-1.3200*

Table (2):The best fitted parameters of equation(5)with ($\chi 2$) value.

Band	$\ell_0(MeV)^{-1}$	$C(MeV)^3$	$E_k(MeV)$	χ^2
G	23.3000	56.0000	0.0000	0.0027
β1	28.0100	56.5000	1.0420	0.0045
β2	35.2000	57.0000	1.4060	0.0276
γ1	8.9999	52.1000	1.1390	0.0870

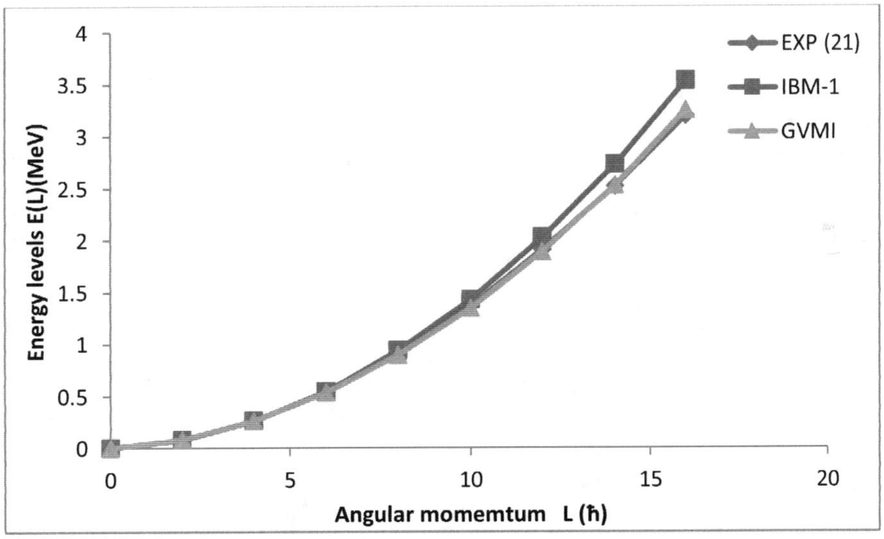

Figure(1): Three calculated IBM-1, GVMI, and experimental[21] E(L)versus L for g-band

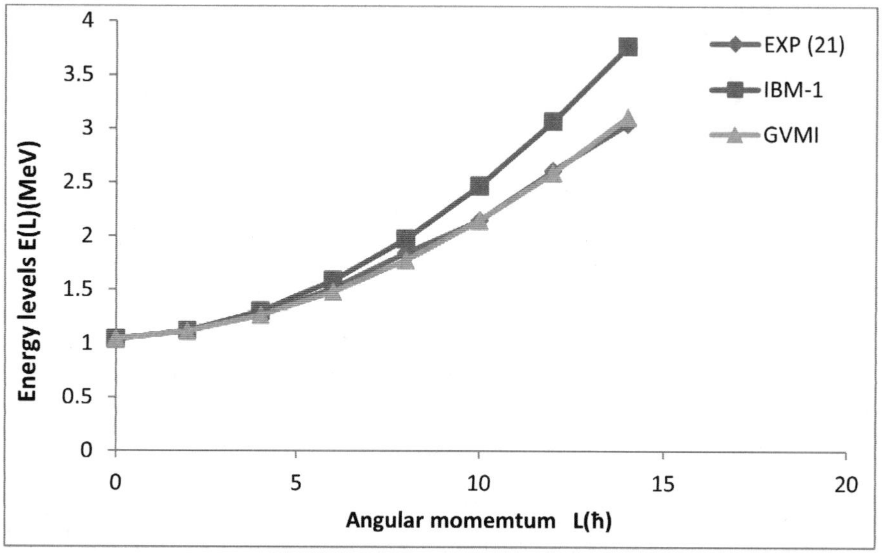

Figure(2): Three calculated IBM-1, GVMI, and experimental[21] E(L)versus L for B1-band

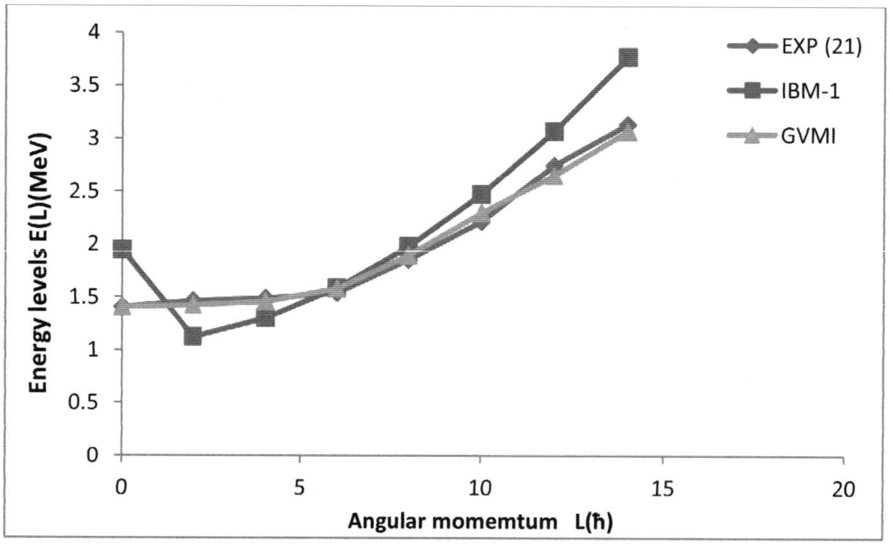

Figure(3): Three calculated IBM-1, GVMI, and experimental[21] E(L)versus L for B2-band

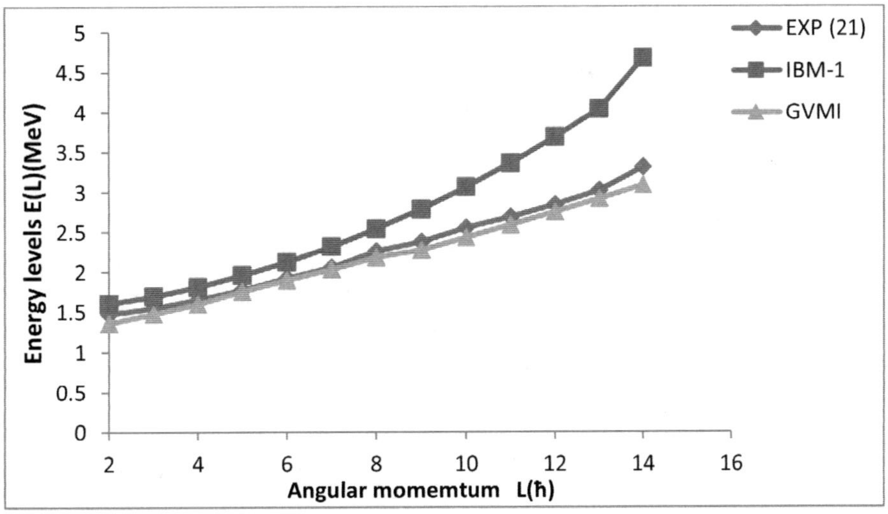

Figure(4): Three calculated IBM-1, GVMI, and experimental[21] E(L)versus L for gamma 1-band

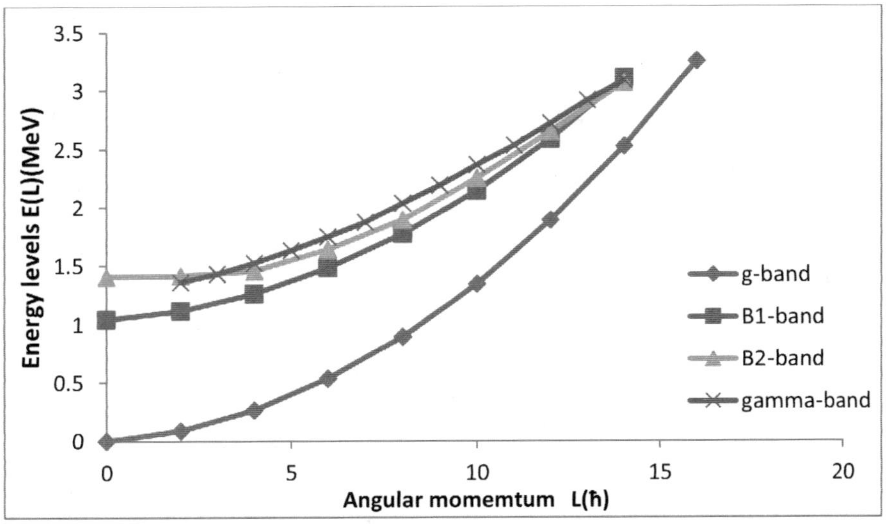

Figure (5): The energy band crossing [E(L) as a function of angular momentum(L)]using the GVMI model. The parameters used for calculations are in Table (2)

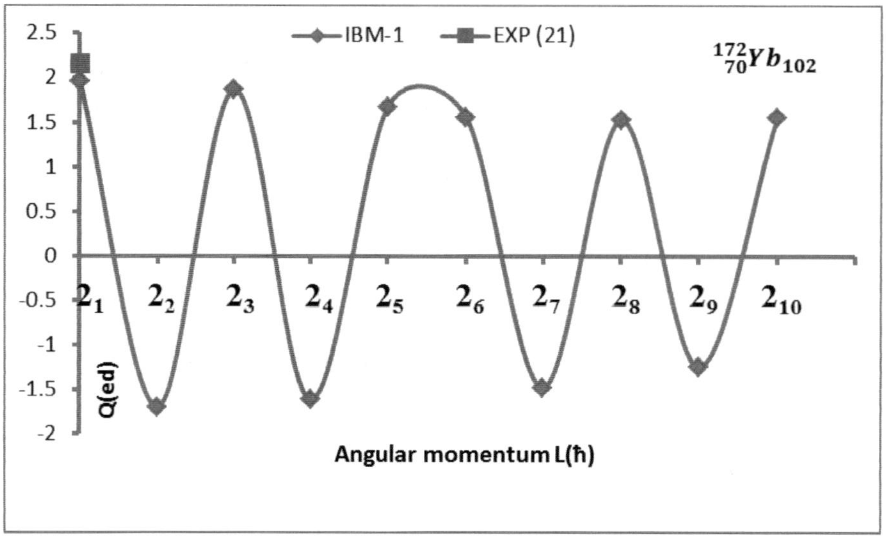

Figure (6): The electric quadrupole moment as a function of angular momentum L=2_1 to 2_{10}.

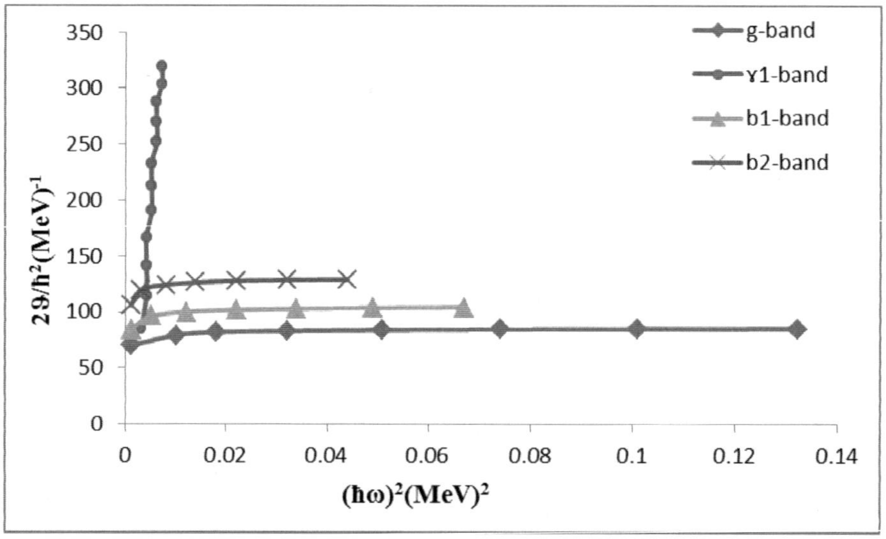

Figure (7): The back bending ($2\vartheta/\hbar^2$) versus $(\hbar\omega)^2$ plot for chosen nuclei using (GVMI) model.

IOP Conf. Series: Materials Science and Engineering 1095 (2021) 012013 doi:10.1088/1757-899X/1095/1/012013

Synthesis of Iron-nickel Particles by Co-precipitation Technique and Used as a Contrast Medium in an MRI Machine

T H Mubarak [1], O A Mahmood[2] and W A Shatti[3]

Department of Physics, University of Diyala, Diyala, Iraq.

E-mail: : dean@sciences.uodiyala.edu.iq [1]olfatahmed183@gmail.com [2] wafaa@sciences.uodiyala.edu.iq [3]

Abstract. In this study, Nickel Zinc Ferrite was prepared using the co- precipitation method according to the chemical formula $Ni_{1-x}Zn_xFe_2O_4$, where the values of (x = 0,0.1,0.3,0.5,0.7,0.9) respectively, the structural and magnetic properties were studied through X-ray diffraction spectroscopy. FESEM showed the porosity and distribution of the homogeneous nanoparticles roughly، also its particle size. The vibration sample Magnetometer (VSM)was used to study magnetic properties, and the Fourier transform infrared (FTIR) spectra were measured in the range of 400-4000 cm^{-1}. Magnetic nanoparticles effect on contrast factors in magnetic resonance imaging (MRI) images was studied, it showed Good and robust imaging technique for diagnosis and post-treatment evaluation for a wide range of diseases. MRI contrast can be enhanced by using negative or positive contrast agents, resulting in clearer (T1-weighted) or darker (T2-weighted) images, respectively, with super-magnetic nanoparticles yielding T2 contrast agents widely used in molecular imaging applications and cellular. As most of the work in this field focuses on the acquisition of biological species using nanoparticles, which in turn will improve the magnetic properties that lead to higher imaging.

Keywords: Co-precipitation method, MRI, Contrast .

1. Introduction

The magnetic properties of nanometer-sized particles have attracted much attention in recent years due to their unique properties. For the preparation of magnetic liquids, nanoparticles with a particle size of the order of 10 nm and a narrow size distribution were required [1]. Magnetic nanoparticles have aroused increasing interest among researchers in various fields due to their wide applications such as information storage system, medical diagnostics, ferrous fluid technology, etc. [2,3]. In this work, the prepared particles were used as a contrast material in MRI as a biological imaging tool with the fact that these particles or tissues have good magnetic sensitivity and when they interact with contrast agents, they provide high contrast images [4]. These prepared magnetic particles have the ability to produce dipoles and magnetic fields with strong spatial differences, which accelerate the spin relaxation of adjacent water molecules in soft tissues and thus determine the spatial resolution and sensitivity of the MRI. It can lead to longitudinal (T1) and transverse (T2) relaxation. [5].

Magnetic nanoparticles (MNP) can accelerate the relaxation of adjacent water molecules by creating local magnetic fields, inside the prepared sample molecules [6]. Magnetic spinel minerals are more efficient, and efficiencies are influenced by the size, composition, atomic structure, and surface characteristics [7]. One of the fundamental aspects of MRI research is the need for high sensitivity, which in turn will allow high-contrast MR images even at a low dose, which results in a lower risk of toxicity of the MRI itself [8]. The sensitivity of the contrast agent T2 increases as the magnetic values of the prepared particles increase. Increasing the size of the magnetic core can lead to the use of strong magnetizing materials and promoting the diffusion of water using low porosity nanostructures associated with the nanoscale size of magnetic particles [9].

2. Experimental Procedure

Ferrite nanoparticles were prepared with the following formula $Ni_{1-X}Zn_XFe_2O_4$ by co-chemical precipitation method according to the proportions (X= 0,0.1, 0.3, 0.5, 0.7, 0.9) to prepare Ni_{1-X} Zn_X Fe_2O_4 (NP) particles from $FeCl_3$ mixtures and $ZnCl_2$ solutions were used. $NiCl_2$, in the presence of 5% NaOH solvent in 100 ml of distilled water at pH = 12 with continuous stirring at 80 ° C, field gradient. In this way the magnetic nanoparticles were prepared and the compatibility of bio-magnetic magnetite was demonstrated. [10] The structural composition of all samples was studied. Using X-ray diffraction (6000 XRD, Cu-kα radiation, λ = 1.54. Shapes for all samples [11]. The particle size was calculated using Scherer's equation (1) The lattice constant (a) was worked out. Calculate it also using Equation (2) [12]. The magnetic properties that can be obtained with (VSM) were studied, after obtaining nano ferrite powders, and based on the structural and magnetic properties, the prepared materials are transformed into a magnetic liquid using deionized water with an organic solvent to ensure the dispersion process. The powders were dissolved using an ultrasound machine (Ultra-sonic) for a period of 5 hours where the nanomagnetic liquid shown in figure 1 was obtained, and it was found that the time factor has no effect on the prepared liquid, and (1) represents the prepared liquid. After one month, the magnetic fluid is thus used in the medical application by using it as a coloring tool in the magnetic resonance device. It was applied to the type of laboratory mice (albino) /pulpec that was purchased from the Center for Cancer Research and Medical Genetics.

$$D = \frac{k\lambda}{\beta_{hkl} cos\theta} \tag{1}$$

where D is the average crystal size, K is the Scherrer coefficient (0. 9), λ is the x-ray wavelength, θ is Bragg's angle (2θ) and β the full width at half-maximum (FWHM) in radians.

$$a = \frac{n\lambda \sqrt{h^2+k^2+l^2}}{d 2 sin\theta} \tag{2}$$

where a the lattice parameter, d is the interplanar distance of each plane and (hkl) are Miller indices, θ is Bragg's angle, λ is the x-ray wavelength, n = 1, 2, 3…..

Figure 1. The stabilized Ferro fluid containing cysteine magnetite nanoparticles after 1 month (concentration: 2 mg Fe / ml)

3. Results and Discussion

The structural properties of the prepared particles have been studied using XRD data, and the average crystal size of the diameter of the crystals can be estimated according to Shearer equation (1). Table 1 shows the particle size, the particle retinal values, and the figure 2 showing the results of X-ray diffraction.

Table 1. XRD analysis of peak position (311) crystal size and Lattice constant.

IOP Conf. Series: Materials Science and Engineering 1095 (2021) 012013 doi:10.1088/1757-899X/1095/1/012013

Sample	Molur ratio	Composition	2Θ	a A⁰	D (nm)
B1	0	$NiFe_2O_4$	35.662	8.341	6.9
B2	0.1	$Ni_{0.9}Zn_{0.1}Fe_2O_4$	35.677	8.337	9.1
B3	0.3	$Ni_{0.7}Zn_{0.3}Fe_2O_4$	35.116	8.467	14.3
B4	0.5	$Ni_{0.5}Zn_{0.5}Fe_2O_4$	35.494	8.381`	13.2
B5	0.7	$Ni_{0.3}Zn_{0.7}Fe_2O_4$	35.417	8.394	7.2
B6	0.9	$Ni_{0.1}Zn_{0.9}Fe_2O_4$	35.327	8.417	14.5

The X-ray diffraction result of ferrite nanoparticles of nickel-zinc imaged with the chemical formula $Ni_{1-x}Zn_xFe_2O_4$ (x = 0, 0.1, 0.3, 0.5, 0.7, 0.9) analyzed by comparing it with the standard JCDPS card. All observed peaks matched well with JCPDS card No. 52-0278 of nickel ferrite, confirming the single-phase cubic spinel Fd m of ferrite samples. From figure (2) it can be seen that the small shift of the peaks towards the short angle to increase the concentration of Zn^{+2}, the X-ray test also showed that there is no growth for stages other than the ferrite phase.

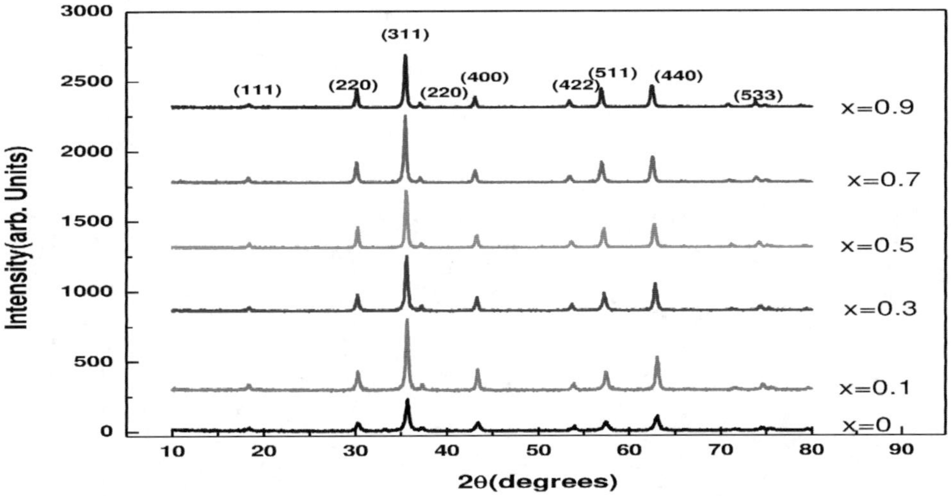

Figure 2: XRD patterns for Nickel zinc Ferrite nanoparticles at 700 °C .

Figure 3 show the electron microscopy of field emissions of ferrite nickel-zinc samples according to the chemical formula $Ni_{1-X}Zn_X$ Fe_2O_4 giving a clear appearance of the composition of the porous nature and nanoparticles and the nearly homogeneous distribution of the size of the nanoparticles. From these images, it can be seen that the micrograph of nanoparticles with spinel structure replacing nickel ferrite reveals that the fine structure of ferrite has been affected by substitution and appears stacked with homogeneous spherical and polyhedral particles, and thus the particle represents the sizes fixed on each image average grain sizes for each A group of closer particles in each sample, then the

IOP Conf. Series: Materials Science and Engineering 1095 (2021) 012013 doi:10.1088/1757-899X/1095/1/012013

clarified particles represent the maximum size, meaning that there is also another size range in another region. The grain size also changes according to the change in the X ratio due to the difference between the radii of zinc and nickel ions.

X = 0.1

X = 0.1

X = 0.3

X = 0.5

X = 0.7

X = 0.9

Figure 3: FE-SEM micrographs of $Ni_{1-x}ZnxFe_2O_4$ nanoparticles for (X=0.0,0.1,0.3,0.5,0.7,0.9) at 700 ^{O}C.

The formation of the porous structure of the ferrite samples may be due to the release of excess amounts of gases during the combustion method [13] In the FE-SEM images, some agglomerated areas appear. This is most likely due to the naturally occurring interaction between magnetic nanoparticles and the thermal process. It was observed in many cases the tendency for the nanoparticles to agglomerate in Nano crystalline spinel ferrites [14]. Infrared Fourier transforms spectra (FTIR) measured in a range of 400-4000 cm^{-1} for prepared samples as shown in Figure (4) Spectra give a composition about the spinal structure and network vibration, while the ranges are around 3846-1417 cm $^{-1}$ attributed to the expansion vibration of the hydroxyl group OH, the ranges at

1384 cm-1 indicate the group of carboxyl's (C=O) the vibration and absorption range is about 1037 cm-1 The nitrate group produces no stretch vibrations [15]. Also, two vibration bars between 600 cm^{-1} and 400 cm^{-1} related to the formation of the spinal structure can be observed according to Waldron's theory all these values noted in the table (2) [16].

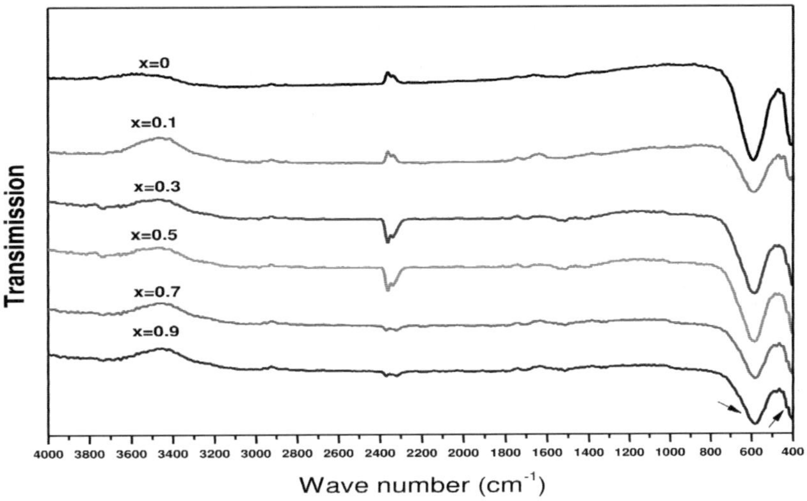

Figure 4. FTIR spectrum of nickel-zinc ferrite, $Ni_{1-x}Zn_xFe_2O_4$ (x = 0.0, 0.1, 0.3, 0.5, 0.7, 0.9).

Table 2: The FTIR absorption bands measured for Ni-Zn spinal ferrite samples .

Sample	Molur ratio	Composition	υ_T cm^{-1}	υ_B cm^{-1}
B1	0	$NiFe_2O_4$	583.6	421.7
B2	0.1	$Ni_{0.9}Zn_{0.1}Fe_2O_4$	574.3	428.0
B3	0.3	$Ni_{0.7}Zn_{0.3}Fe_2O_4$	586.8	421.7
B4	0.5	$Ni_{0.5}Zn_{0.5}Fe_2O_4$	593.0	415.5
B5	0.7	$Ni_{0.3}Zn_{0.7}Fe_2O_4$	580.6	428.0
B6	0.9	$Ni_{0.1}Zn_{0.9}Fe_2O_4$	574.3	402.3

The magnetic properties of the prepared particles were analyzed using a magnetic scale (VSM) at room temperature. figure 5 shows the M-H curves for the samples. The ring is narrow which means that prepare samples are soft magnetic material. The values of saturation magnetization (Ms) and coercive (HC) were extracted directly from these curves as in the figures and were included in the various Zn content in table 3. It can be seen from figure 5, where the saturation and Compulsive magnetization increases with zinc content and then decreases to (x= 0.5). The behavior of the saturation magnetization may be explained by the distribution of actions and by the superior exchange interactions, and also because of the different ratios of magnetic moments in every magnetic domains according to the changing of (x) ratio for every sample.

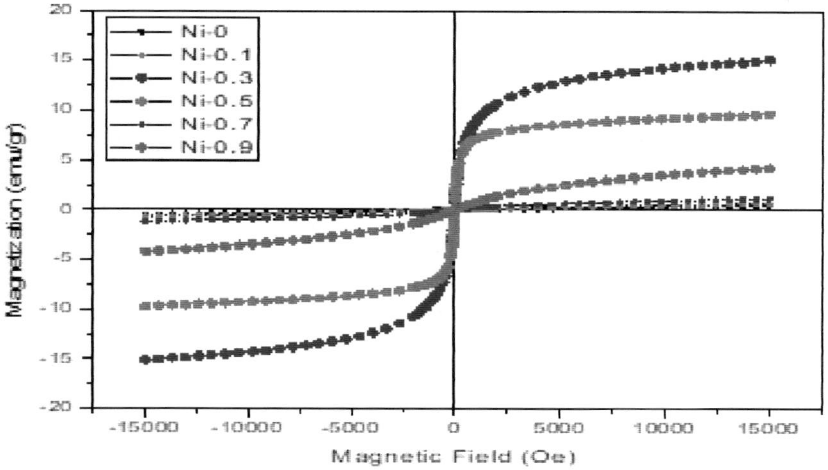

Figure 5. Magnetic hysteresis loops of $Ni_{1-x}Zn_xFe_2O_4$ (x=0, 0.1, 0.3, 0.5, 0.7 and 0.9).

Table 3. The values of saturation magnetization, remanence, connectivity, squareness ratio Mr/Ms, anisotropy constant, and permeability.

Sample	Molur ratio	Composition	M_S (emu/g)	M_r (emu/g)	H_C (Oe)	$S=M_r/M_S$
B1	0	$NiFe_2O_4$	0.54	0.004	21.27	0.007
B2	0.1	$Ni_{0.9}Zn_{0.1}Fe_2O_4$	0.691	0.03	80.54	0.04
B3	0.3	$Ni_{0.7}Zn_{0.3}Fe_2O_4$	15.2	4.627	27.21	0.29
B4	0.5	$Ni_{0.5}Zn_{0.5}Fe_2O_4$	9.68	`4.728	154.4	0.48
B5	0.7	$Ni_{0.3}Zn_{0.7}Fe_2O_4$	1.177	0.08	89.42	0.06
B6	0.9	$Ni_{0.1}Zn_{0.9}Fe_2O_4$	4.218	0.301	229.4	0.07

4. Medical Application

For the MRI study, samples with nanoparticles and magnetic properties present in the prepared particles were selected within the B3B4 samples. Laboratory rats were used for an MRI scan, and the mice were anesthetized and then the stable nanoparticles were introduced through aqueous ferric fluid prepared at a dose of 2.5 mg (Fe) / kg bw. An MRI scan was performed half an hour after injection [17]. We observe the accumulation of macromolecules and aggregate mainly in tissues such as liver and spleen, however, smaller (28 nm) will penetrate the tissue. From the injected mice, the mice were anesthetized laterally in the groin area in an MRI chamber, and the side of the mouse was placed for a sagittal plane, then T1 and T2 MRI was performed before the nanomaterial was injected and then compared with those injected in figure 6. The magnetic Nanofluid was injected with Ferrite Nickel Zinc, sample B3 with a concentration of X = 0.3 Nanoscale (14.3nm). The examination with the MRI the device of this sample revealed pictures of the T1 promoter shown in figure 7, which shows a highly sensitive appearance of the material and shows the penetration of the material to the tissue at Likely T2 shown in figure 8.

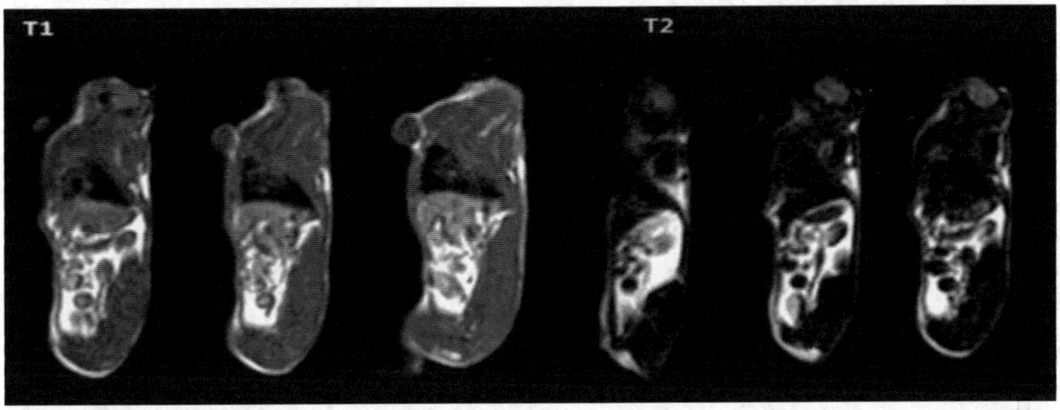

Figure 6. shows the resulting images before using the magnetic nanoparticles.

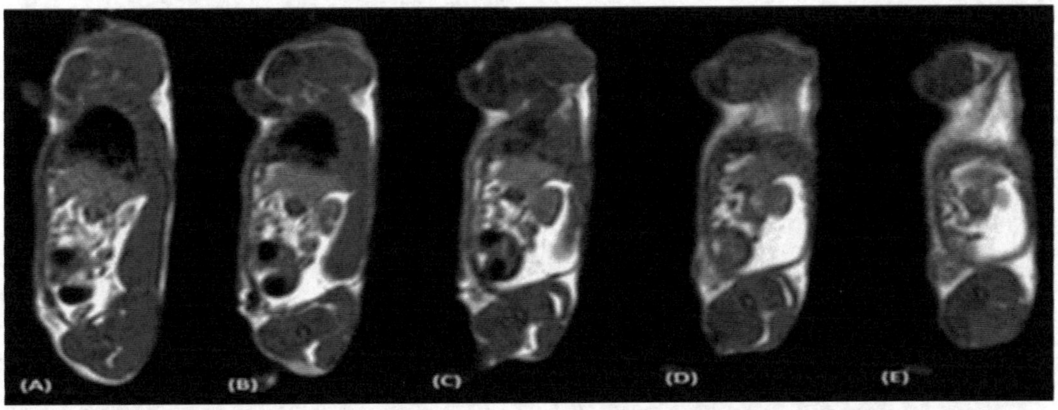

Figure 7. T1 shows the Images from the B3 sample MRI that show clarity of the images produced.

IOP Conf. Series: Materials Science and Engineering 1095 (2021) 012013 doi:10.1088/1757-899X/1095/1/012013

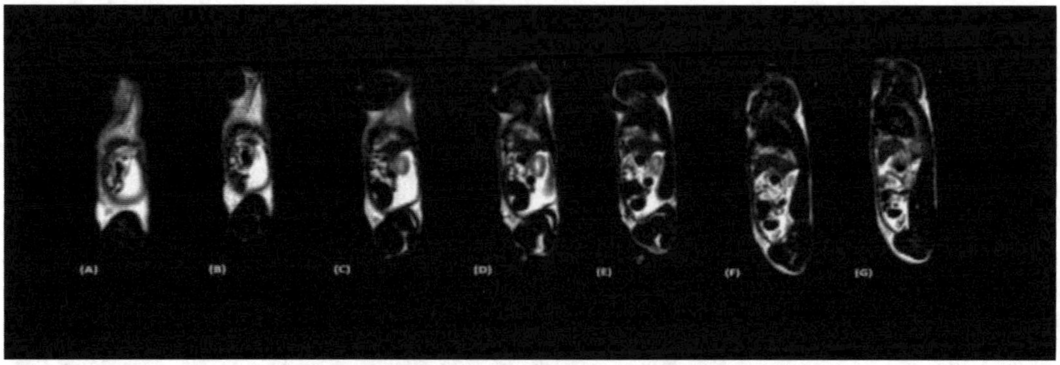

Figure 8. T2 shows the Images from the B3 sample MRI that show clarity of the images produced

Weighted MRI scan. B4: X = 0.5. An injection in the right revealed a hypersensitive area also in both T1 and T2 MRI images, however, the fluid appeared to penetrate deeper than the previous material (figure 9 to 10). It was injected, showing good resolution images [18].

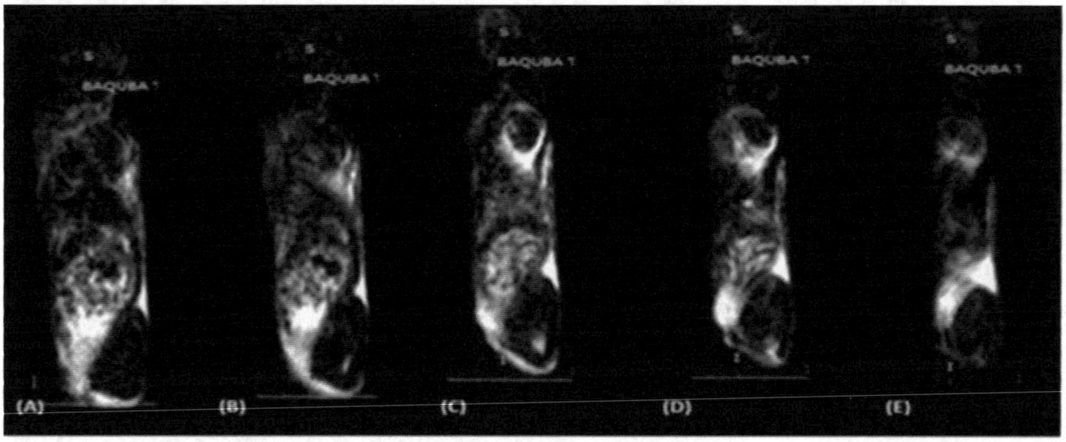

Figure 9: A ferromagnetic fluid present at the intestine (*) was shown when the T1.

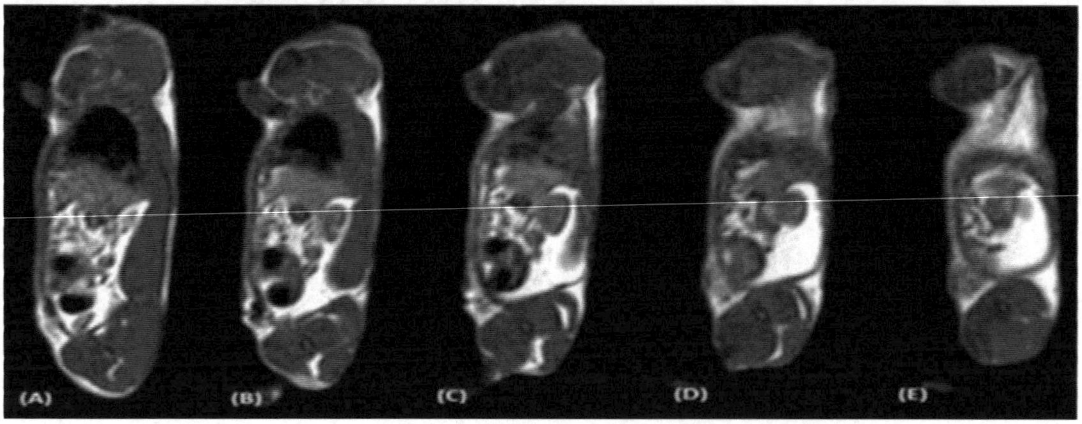

Figure 10: Shows that the fluid had a greater penetration than the previous material. There were no statistically significant differences between the previous liquid and ferromagneti

5. Conclusions

Stable ferromagnetic nickel particles were fabricated with average nano scale size using co-chemical precipitation method. The nano scale liquids were prepared using an Ultra-sonic machine. The prepared liquids were marked by an iron nanometer stability. This was used as a staining agent in studying images in the MRI machine, and it has been observed to enhance good contrast in post-injection MRI.

References

[1] Rosensweig, R. E. and Rosensweig, R. E., 2013. Ferrohydrodynamics. Courier Corporation. Courier Corporation, pp.46-50;
[2] Kim, Y. I., Kim, D. and Lee, C. S., 2003. Synthesis and characterization of CoFe2O4 magnetic nanoparticles prepared by temperature-controlled coprecipitation method. Physica B: Condensed Matter, 337(1-4), pp.42-5;
[3] Pathmamanoharan, C. and Philipse, A.P., 1998. Preparation and properties of monodisperse magnetic cobalt colloids grafted with polyisobutene. Journal of colloid and interface science, 205(2), pp.340-353;
[4] Xu, W., Kattel, K., Park, J.Y., Chang, Y., Kim, T. J. and Lee, G. H., 2012. Paramagnetic nanoparticle T 1 and T 2 MRI contrast agents. Physical Chemistry Chemical Physics, 14(37), pp.12687-12700;
[5] Geraldes, C. F. and Laurent, S., 2009. Classification and basic properties of contrast agents for magnetic resonance imaging. Contrast media & molecular imaging, 4(1), pp.1-23;
[6] Prausnitz, M. R. and Langer, R., 2008. Transdermal drug delivery. Nature biotechnology, 26(11), pp.1261-1268;
[7] Lee, N. and Hyeon, T., 2012. Designed synthesis of uniformly sized iron oxide nanoparticles for efficient magnetic resonance imaging contrast agents. Chemical Society Reviews, 41(7), pp.2575-2589;
[8] Brooks, R. A., Moiny, F. and Gillis, P., 2001. On T2-shortening by weakly magnetized particles: the chemical exchange model. Magnetic Resonance in Medicine: An Official Journal of the International Society for Magnetic Resonance in Medicine, 45(6), pp.1014-1020;
[9] Lee, H., Yoon, T. J., Figueiredo, J. L., Swirski, F. K. and Weissleder, R., 2009. Rapid detection and profiling of cancer cells in fine-needle aspirates. Proceedings of the National Academy of Sciences, 106(30), pp.12459-12464.;
[10] Ursache-Oprisan, M., Foca-nici, E., Cirlescu, A., Caltun, O. and Creanga, D., 2010, December. Oleate coated magnetic cores based on magnetite, Zn ferrite and Co ferrite nanoparticles—preparation, physical characterization and biological impact on helianthus annuus photosynthesis. In AIP Conference Proceedings (Vol. 1311, No. 1, pp. 425-430). American Institute of Physics;
[11] Kareem, M. M., Khodair, Z. T. and Mohammed, F.Y., 2020. Effect of annealing temperature on Structural, morphological and optical properties of ZnO nanorod thin films prepared by hydrothermal method. Journal of Ovonic Research Vol, 16(1), pp.53-61;
[12] Khodair, Z.T., Mohammad, A.M. and Khadom, A.A., 2020. Investigations of structural and magnetic properties of Cu1-xVxO nanostructures prepared by sol-gel method. Chemical Data Collections, 25, p.100315;
[13] Gowreesan, S. and Kumar, A. R., 2017. Effects of Mg 2^{+} ion substitution on the structural and electric studies of spinel structure of Co 1− x Mg x Fe 2 O 4. Journal of Materials Science: Materials in Electronics, 28(6), pp.4553-4564;

[14] Naseri, M.G., Saion, E.B., Ahangar, H.A. and Shaari, A.H., 2013. Fabrication, characterization, and magnetic properties of copper ferrite nanoparticles prepared by a simple, thermal-treatment method. Materials Research Bulletin, 48(4), pp.1439-1446;

[15] Singh, R.K., Upadhyay, C., Layek, S. and Yadav, A., 2010. Cation distribution of Ni 0.5 Zn 0.5 Fe 2 O 4 nanoparticles. International Journal of Engineering, Science and Technology, 2(8);

[16] Shahane, G.S., Kumar, A., Pant, R.P. and Lal, K., 2010, October. Structural and Magnetic Properties of Ni-Zn Ferrite Nanoparticles. In AIP Conference Proceedings (Vol. 1276, No. 1, pp. 374-380). American Institute of Physics;

[17] LaConte, L., Nitin, N. and Bao, G., 2005. Magnetic nanoparticle probes. Materials Today, 8(5), pp.32-38;

[18] Ahmad, T., Rhee, I., Hong, S., Chang, Y. and Lee, J., 2011. Ni-Fe2O4 nanoparticles as contrast agents for magnetic resonance imaging. Journal of nanoscience and nanotechnology, 11(7), pp.5645-5650;

The Role of Teaching Methods Based e-learning Technologies in Iraq Higher Education Institutes

Etmad N.Fayadhm[1], Bassim M. Salih[2] and Omer K. Jasim Mohammed[3]

[1]Continuing Education Center، University of Fallujah
[2]Computer Center، University of Fallujah
[3]Department of Quality Assurance and Academic Accreditation، University of Fallujah

[1]dr.etmad.naji@uofallujah.edu.iq, [2]bassim.m.s1977@uofallujah.edu.iq, [3]omerk.jasim@uofallujah.edu.iq

Abstract: Currently, the world witnesses a major revolution in the world of communications and information technology, where technology permeated all aspects of social, learning and military life in both phases: public and private. This paper measures the rate of using teaching methods of e-learning technology in higher education Iraqi institutes; university of Fallujah is a sample of such institute. This paper used the descriptive analytical approach (DAA) in collecting, classifying gathering information for five colleges by taking 30 measurement coefficients that are compatible with teaching operation. The obtained results demonstrate that a number of lecturers and students have used the e-learning applications in the class management are extremely low rates. Finally, this study has resulted in a novel concept, usually called "change –resistance".
Keywords: E-learning, teaching methods, traditional education, Iraqi universities, higher education.

1. Introduction

With the rapid and growing development of Internet communication Technology (ICT), educational institutions have revised their goals and practices. Accordingly, all universities looked for the most suitable methods and the best styles through which they can provide educational experiences to their students, instead of the methods adopted on memorization and indoctrination.

New trends in technology led to the emergence of new patterns of learning and education and new teaching methods using the techniques of e-learning, which in turn increase the knowledge rate and support traditional education. So, universities have sought to continue the scientific plans for the availability of an interactive learning environment, which further entrenches the concept of individual or subjective education. The learner continues to learn according to his capacity, ability and speed of learning and according to his previous experience and skills. E-learning is one such evolving pattern of so-called general distance learning and computer-based education in particular[7]

Generally, university members believed and considered that the traditional education style (face-to-face) and classroom interaction as effective and useful method, according to 95% of teaching staff believe that the traditional education style based face to face style is still the most effective way and they use it tends to start with less complex activities [2].

E-Learning is new emerging phenomena of information technology, internet family and computing, it presents a technical achievement for the education sector and adds smoothly technical roles in the lecture management, deliver, discussion and examination. This paper presents a scientific philosophy view of the effecting of e-learning in the universities and the educational process. Also, this paper classified and analyze gathered information using the descriptive analytical approach (DAA) [16], for five colleges by taking 30 measurement coefficients that are compatible with the teaching operation in the university of Fallujah. This context provides a holistically view for the role of e-learning in the Iraqi higher institutes. The student has become the focus of the educational process such as individual learning, programmed learning and self-learning. E-learning has become a reality and the need for it urgently needs. Therefore, universities must improve its employment and incorporate it into teaching methods [3].

The rest of the paper is organized as following. Section 2 presents the methodology and objectives of the study; the concept of E-Learning is featured in Section 3. Section 4 shows the E-Learning elements, while, E-learning patterns and tools are depicted in Section 5. Section 6 illustrates the research community of the paper. Statistical Trend is showed in Section 7. Section 8 presents a result and analysis. Finally, conclusions and future work are illustrated in Section 9.

2. Methodology and Objective

This study adopts the descriptive analytical approach (DAA) that is describing the collected data in quantity and dynamic scalable style for the gathered information. DAA is a statistical method that is used to search and summarize historical data in order to identify patterns or meaning and its completely depends on two basic techniques, data aggregation and data mining. These techniques are used in descriptive analytics to discover historical data and activity pattern. Data is first gathered and sorted by data aggregation in order to make the datasets more manageable by analysts. While, the data mining describes the next step of the analysis and involves a search of the data to identify patterns and meaning. Identified patterns are analyzed to discover the specific ways that learners interacted with the learning content and within the learning environment [16].

DAA helps the author to reveal the relationship between the user's activities and data collected to interpret them. Additionally, the main objectives of this study listed as below: -

1- To identify the challenges of e-learning which are faced by faculty members at the University of Fallujah in the practice of e-learning.
2- To provide feedback to university leaders and higher education policymakers.
3- To help the university leader address all obstacles in order to successfully implement the e-learning system.
4- To identify which faculty members use the e-learning system and their attitudes towards its use.
5- Supporting traditional education through e-learning.

3. E-Learning Concept

E-Learning is a modern method of teaching that is employing a progressive communication mechanism from computers, networks, multimedia, graphics, electronic libraries, as well as Internet portals, whether remotely or in the classroom [8]

E-learning depends primarily on computers and networks to deliver knowledge and skills. E-learning is characterized by flexibility and positive adaptation to events to achieve the desired objectives and it becomes a reality and the need is dire. Accordingly, all universities should improve their staff in order to convertibly integrating into new teaching methods. A lot of features associated with adoption of e learning technology, like following [9]:-

i. Providing interactive educational content with explanations, exercises, interaction and follow-up in a partial or comprehensive manner in the classroom or remotely through advanced programs stored in the computer or over the Internet.
ii. Providing an integration educational system that uses technologies to strengthen and expand the educational process.
iii. Providing a new method for the lectures delivering and explanation based modern techniques of the computer and the World Wide Web and their multimedia, such as:

Software and CDs, educational, dialogue and discussion arenas, e-mail, dialogue and discussion arenas.

3.1 E-Learning Elements

The educational environment for e-learning consists of the following elements [4]; [11]:

i. Instructor: the most important element of e-learning technology and serves a source of education, also, the teacher should have a good knowledge of computer use and technology like Internet and e-mail. Furthermore, a teacher should be having the ability to teach using modern teaching techniques or open learning systems like Moodle, Google for education, etc.
ii. Learner: learner receives through the practice and self-research (i.e.) develops a self-learning skill. The learner should be hearing about a group of students and interacts with others in a manner independent of them, like as an instructor the learner must have an experience in the computer usage, Internet and e-mail.
iii. Computer Experts: a group of specialized persons in computer, software, Internet components, and provide help in creation and design steps.

3.2 E-Learning Patterns and Tools

This section explains the main idea of E-learning: patterns and categories. Generally, E-learning divided into three main patterns, synchronous, Asynchronous, and Blended e-learning. Synchronous e-learning (Simultaneous) is the type of education in which the teacher and the learner are required to be present in the same educational environment and at the same time and place, with simultaneous contact with text, voice, and video or electronic encounters [6]; [12].

Chat engine, audio conferences, video conferencing (video conferences), white board, and satellite programs (satellite programs) are well-known tools of the synchronous pattern.

While, a synchronous e-Learning in which education is indirect since the teacher and the learner are not required to be present in the educational environment at time and the student is often not obliged to have a certain time or place and the teacher can put the content with a teaching plan and a calendar on the site tutorial, then enters the learner of the Web site at any time. A lot of examples of this pattern such as E-mail, Textile Network (World wide Web), Mailing lists (Mailing list), Discussion groups (Discussion Groups), Transfer files (file Exchange), Interactive video (Interactive video), and CD-ROMs [13].

The integrated education (Blended learning) is the type of education defined as a mixture of traditional and e-learning, and used the e-learning tools as part of traditional education.

In addition, the sources of e-learning evolve day by day including: electronic books, databases, Encyclopedias, periodicals, educational sites, E-mail, digital libraries, virtual classroom, video conferences, Self-education, and online educational (sites Internet Sites).

4. Research Community

This paper focuses on the University of Fallujah as a sample of higher education Iraqi institute, it consists of 5 colleges: college of Medicine, college of Veterinary Medicine, college of law, college of Islamic Sciences, and college of Administration and Economics. With relevance to human resources, 210 as a teaching staff, this study takes 50% as a sample. The measurement process completely depends on five basic steps that are preparation, honesty, rearrangement, guidelines, and persistent steps, see Fig. 1.

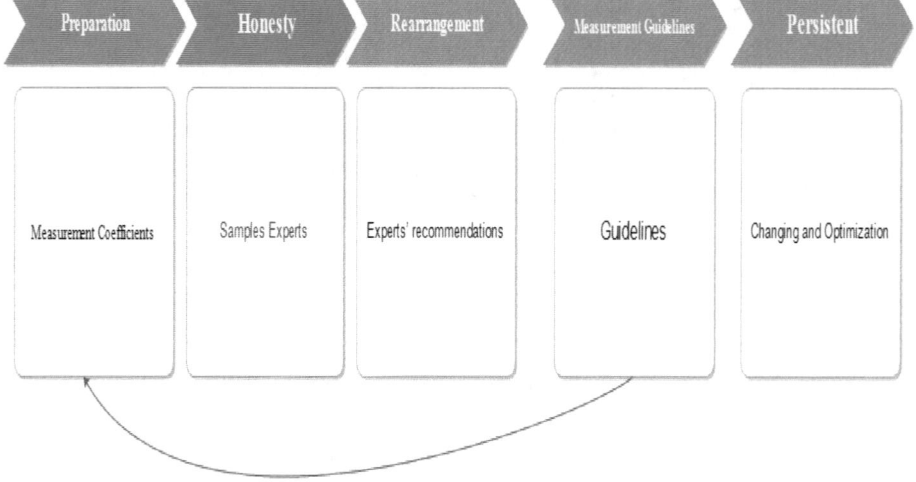

Fig. 1 Measurement Process Steps

1. **Preparation**: generally, this paper depends on the [11], research, after optimized it by adding some criteria and increase it to 35 measurement coefficients.
2. **Honesty:** in this step, author divided the mentioned 35 measurement coefficients in to five-tiered scales of approval (5, 4, 3, 2, 1). This classification includes of a group of experts and specialists from the *Universities of Fallujah* and *University of Anbar* in psychology, education philosophy, and teaching methods.
   The selection process is very complex and accurate. Usually, it measures the rate of suitability coefficient or not and check if its need to be modified or not. Consequently, about 80% of the mediators fully agreed and adopted such coefficient with new optimization criteria. According to our questionnaire, this paper reflects that the honesty step is more important step, due to show the appropriation and agreement between the test samples of people and specialists.
3. **Rearrangement:** regarding opinions and suggestions of the nominated experts, author followed and adopted 80% from the experts' recommendations that are related to measurement coefficients. So, according to [10] this paper follows the right methodology and analysis.
4. **Measurement Guidelines:** the guidelines have been prepared for the scale of the samples answer, and what is meant according to the study survey, especially in preparation step to produce the accurate results.
5. **Persistent:** author going to change and optimize the research sample elements (teacher and students), also, this step explains the clarity and originality of measurement coefficients. Then, author returned a new modification to the experts and teachers to find the correlation coefficient. Furthermore, author test the measurement coefficients with 20 teachers (research community sample), then, explores the formulation and new

guidelines after some minor modifications. Finally, author deploy a standard mining tool in order to use with 30 optimal coefficients for a five-tiered scale of approval).

5. Statistical Trend

In this stage, the scale has been distributed to the research sample and calculation the final score for each coefficient in the theory and weighted mean. The computation of weight has been achieving by the following equations:

$$TM = \frac{atw}{an} \quad (1)$$

$$WTM = aw \times \frac{ai}{sn} \quad (2)$$

$$TC = \frac{wc}{hw} \times 100 \quad (3)$$

Where:

TM : Theoretical mean, atw: alternatives total weights; an= alternatives number; WTM= Weights Theoretical mean; aw =alternative each weight; ai = alternative each iteration; sn= sample number of individuals; TC= Theoretical Center; wc= Weighted center; hw= higher weight.

According to eq. (1,2, and 3), each coefficient has WTM equal to (3) or greater than. Thus, this is one of the reasons why e-learning applications are not practiced from the university's point of view. The following section (Sec.6) shows the analytical analysis for the obtained results.

6. Results and Discussions

This section shows the obtained result after received the answer from the selected persons (community samples). The questionnaires show that the only 10% of the college members used the E-learning applications. Table 1 shows the value of the weighted mean and weighted center for each measurement coefficient (9-coefficients) and shows the order in descending style. This computation has been made on administrative departments.

Table 1: Weighted mean and weighted center of the measurement coefficient on the administrative side

NO.	Measurement Coefficient	WM	WC
1	Unavailability of computers.	4.465	89.3
2	Problems related to the availability of electric power.	4.431	88.62
3	Lack of room equipment with the necessary tools for e-learning practice.	4.428	88.56
4	Non-training of faculty members to use e-learning applications.	4.396	87.92
5	The mainstream educational system does not allow the use of e-learning	4.362	87.24
6	Lack of material resources for e-learning programs	4.328	86.56
7	Internet vulnerability and interruptions.	4.324	86.48
8	Not suitable for the halls environment when introducing any	4.30	86

	technological educational method.		
9	Lack of cooperation between universities in the exchange of experiences in the field of e-learning	3.45	69

As showed in Table 1, the type of administrative style is directly effects of the delay or prevent the practice implementation of the e-learning in the teaching operation. In addition, change – resistance is the one of big challenges with current administrative staff in the university.
With relevance to the teaching and the student samples, Table 2 shows the value of the weighted mean for each measurement coefficient (13-coefficients)- teaching aspects, also, it shows the order in descending style for weighted mean and weighted center.

Table 2: Weighted mean and weighted center of the measurement coefficient on the teacher and student side

No	Measurement Coefficient	WM	WC
1	The teaching feeling that e-learning reduces the power and control of the educational process.	3.34	66.8
2	E-learning is an additional burden on the staff's assigned workload	3.28	65.6
3	Increases the cost of education to the university.	3.25	65
4	The feeling that e-learning lacks confidentiality and security for content and exams.	3.22	64.4
5	Low degree of interaction and social coexistence between the student and the teacher.	3.18	63.6
6	Not to be convinced of the importance of using e-learning in teaching courses.	3.16	63.2
7	Lack of capacity and efficiency in the use of e-learning by students.	3.12	62.4
8	Lack of capacity and competence of the teaching staff to use the English language	3.11	62.2
9	Lack of information and technological skills required for e-learning.	3.9	78
10	Students do not respond to the new style of learning.	2.88	57.6
11	Lack of teaching staff who are proficient in the technological skills required for e-learning.	2.79	55.8
12	E-Learning weakens students ' faith in the educational attitudes and values that the university is working to give them.	2.67	53.4
13	Poor students ' ability to differentiate between what is better to receive and what does not improve their re reception.	2.54	50.8

Table 2 shows that the measurement coefficient of the aspects of teaching and student is significantly similar to administrative side expectation of four coefficients, (19, 20, 21,22). Moreover, Table 3 shows the value of the weighted mean for each measurement coefficient (13-coefficients)- and focuses on the e-learning criteria.

Table 3: Weighted mean and weighted center of the measurement coefficient on the e-learning

No	Measurement Coefficient	WM	WC
1	Difficult to apply in some materials that need practical skills.	4.41	88.2
2	High cost of preparing good software in e-learning style.	3.68	73.6
3	Scarcity of specialists in the design of e-learning learning materials	3.54	70.8
4	Its applications are not available in Arabic.	3.44	68.8
5	Lack of interaction and direct contact between the teacher and the learner.	3.24	64.8
6	Lack of incentives (moral or material) necessary for the e-learning environment.)	3.20	64
7	Does not focus on all senses, but on the sense of hearing and sight.	2.24	44.8
8	The ambiguity of the philosophy of e-learning and its objectives.	2.15	43

As shown in Table 3 the measurement coefficients of the aspects of e-learning is significantly similar to administrative side and teaching side, expect of two coefficients, the results are consistent with the results of many similar studies such as [14] as well as the study [13]; [3], which addressed the role of e-learning in the development of traditional education and raise the level of students and professors on Both.

7. Conclusions and Future Work

From the experience of the authors and their work as a faculty member in the university of Fallujah and use e-learning applications in the teaching operation. This paper focus on the harmony of using the e-learning technology in the university colleges comparing to tradition way. This paper used analytical descriptive approach (ADA) in collecting, classifying the gathered information, and analyzing it. Also, this study takes 30 measurement coefficients that are compatible with teaching operation in the university of Fallujah. The obtained results show that the low ratio of university lecturers and students utilize the e-learning applications in the lecture management and presentation. Finally, this study explores the concept, usually called "change –resistance" and depicts the obtained result compatibles with a lot of surveyed studies like [5]; [13]; [12].

At the future, the author going to deploy the motivational guide for all university staffs to use the e-learning applications. Furthermore, recommend the quality assurance departments in Iraqi universities to determine a score in performance evaluation for all lectures.

Acknowledgment

The authors would like to thank the anonymous reviewers for their valuable comments and suggestions that improved the presentation of this paper and for university of Fallujah in Iraq.

IOP Conf. Series: Materials Science and Engineering 1095 (2021) 012014 doi:10.1088/1757-899X/1095/1/012014

References

[1]Al kudah Khaled Youssef, and Bassam Interview (2013), E-*learning Challenges Facing Faculty Members at Private Jordanian Universities*, Al-Manara Magazine, Vol. 19, No. 3.

[2]Amer, Tariq Abdul Raouf (2007). Distance learning and open education. Amman: Dar Al Yazouri for Publishing & Distribution

[3]Al-Mubarak, Ahmed bin Abdul Aziz (2014), *The Effect of Teaching Using Virtual Classes via the Internet on the Achievement of Students of the College of Education in Teaching and Communication Technologies at King Saud University*, Master Thesis, Riyadh, KSA.

[4]Mohammed Mansur Ibrahim and Muesser Nat (2019), *Blended learning motivation model for instructors in higher education institutions*, International Journal of Educational Technology in Higher Education volume 16, Article number: 12.

[5]Aqeel Thamer Jawad, Chasib Hasan Abboodi (2018), *USE EASYCLASS PLATFORM IN HIGHER EDUCATION: UNIVERSITY OF DIYALA COLLEGE OF BASIC EDUCATION CASE*. International Journal of Computer Engineering & Technology, Vol.4.

[6]Sarah Guri-Rosenblit (2018), E-*Teaching in Higher Education: An Essential Prerequisite for E-Learning*, Journal of New Approaches in Educational Research, Vol 7, No 2 .

[7]Ahmed Hadi Ali, Omer K. Jasim Mohammad, (2019), *Impacting of the E- Platforms on the 4.0th Industrial Educational Revolution,* Proceeding 2019 Proceedings of the 9th International Conference on Information Systems and Technologies, Article No. 29, Cairo, Egypt

[8]Al-Todari, Awad bin Hussein Mohammed (2004), *The electronic school and modern roles of the teacher*, Riyadh, Al-Rushd Library.

[9]Mousa, Abdullah bin Abdul Aziz (2009). E-*Learning Concept, Characteristics, Benefits, Obstacles, Worksheet Introduction to Future School Symposium*, King Saud University College of Education, KSA.

[10]Benjamin S. Bloom, George F. Madaus, J. Thomas Hastings (1983), *Evaluation to improve learning*, New York : McGraw-Hill.

[11]Hamza, Janan Mirza, (2015), *Problems of using e-learning in teaching historical courses from the perspective of teachers*, Journal of the College of Basic Education for Educational and Human Sciences / University of Babylon, No. 20.

[12]KOZÁK, Štefan., (2012), *Advanced control engineering methods in modern technological applications*. In: Proceedings of the 13th International Carpathian Control Conference (ICCC). p. 392-397.

[13]Jinan Mirza Hamza, (2015), *Problems of Electronic Teaching in the Historical Materials from the University Teachers' Point of View*, Journal of Basic Education College Magazine For Educational and Humanities, Issue: 20, pp.: 110-123

[14]Astita Dalal, Sirhan and Omar (2007), *Education and e-learning technology*, Amman: Wael Publishing House.

[15]Judges Khaled Yusuf and Bassam (2013), *E-Learning challenges facing faculty members at Jordanian private universities*, Al-Manara Magazine, vol. 19, No. 3.

[16]URL: https://www.valamis.com/hub/descriptive-analytics (access date 2019).

AUTHOR INDEX